The Economics of Kentucky Coal

Curtis E. Harvey

The Economics of Kentucky Coal

The University Press of Kentucky

ISBN 978-0-8131-5148-9

Library of Congress Catalog Card Number: 76-51160

Copyright © 1977 by The University Press of Kentucky

A statewide cooperative scholarly publishing agency serving Berea College, Centre College of Kentucky, Eastern Kentucky University, The Filson Club, Georgetown College, Kentucky Historical Society, Kentucky State University, Morehead State University, Murray State University, Northern Kentucky University, Transylvania University, University of Kentucky, University of Louisville, and Western Kentucky University.

Editorial and Sales Offices: Lexington, Kentucky 40506

To Judith, Kristen, Mark, and Kathleen

Contents

Preface   xvii
1 The Demand and Supply of Kentucky Coal   1
2 The Eastern Kentucky Coal Industry   17
3 The Western Kentucky Coal Industry   95
4 The Future Use of Energy Resources   151
Notes   159
Bibliography   165
Index   169

Tables

1. U.S. and Kentucky Coal Production, 1960-1975 2
2. Annual Production of the U.S. Bituminous Coal and Lignite Industry, 1969-1975 7
3. Largest Commercial Bituminous Coal Producers in the U.S. during 1973 8
4. Concentration Ratios for Total Bituminous Coal Production in the U.S. for the Years 1950, 1960, 1970, and 1974 9
5. Production of Bituminous Coal in the Appalachian Basin, 1974 20
6. Sulfur, Ash, and BTU Content of Tipple and Delivered Samples of Bituminous Coal 21
7. Distribution of U.S. Bituminous Coal Production by Major User, 1948 and 1973 22
8. The Energy/GNP Ratio, 1947-1970 23
9. National Power Survey Forecast of Total Net Electricity Generation, 1980 and 1985 25
10. Forecast of Fossil Fuels Consumed in Electricity Generation, 1970-1985 26
11. Steam-Electric Plant Fuel Consumption by Region, 1972 28
12. Percentage of BTUs Derived from Coal and Oil by Electric Utilities along the Eastern Seaboard, 1966 and 1972 30
13. Distribution of Bituminous Coal Shipments for Eastern Kentucky and the Remainder of District 8, by Major User, 1970 and 1972 31
14. Shipments of Coal from Eastern Kentucky to Electric Utility Plants, 1972 32
15. Rank-Size Distribution of Electric Utility Market for Eastern Kentucky Coal, 1972 35
16. 1980 CRA Forecasts of Coal Consumption by Electric Utilities and Resulting Eastern Kentucky Demand Forecasts 36
17. Relative Share of Coking Coal Production, 1970 and 1972 38
18. Forecasts of 1980 Domestic Coking Coal Shipments from the U.S. and Eastern Kentucky 39

19. Coal Exports from the U.S. and District 8, 1972   41
20. 1980 CRA Forecasts of Coal Exports from the U.S. and Resulting Eastern Kentucky Forecasts   42
21. Actual 1972 Shipments and Forecast of 1980 Shipments of Coal from Eastern Kentucky   43
22. Coal Output for Four Appalachian Coal Basin States, 1969-1974   44
23. Production of Surface and Underground Mined Coal in West Virginia, Eastern Kentucky, Pennsylvania, and Ohio, 1969-1974   44
24. Production of Coal in Four Appalachian Coal Basin States, Underground and Surface, 1969-1974   45
25. Percentage Distribution of Coal Mining by Type of Mine for Four Appalachian Coal Basin States, 1969-1974   46
26. Coal Characteristics in Four Appalachian Coal Basin States   47
27. Demonstrated Bituminous Coal Reserve Base for Four Appalachian Coal Basin States, January 1, 1974   48
28. Estimated Life of Demonstrated Underground and Surface Coal Reserves in Four Appalachian Coal Basin States at 1974 Production Rates   49
29. Surface Mine Production in Four Appalachian Coal Basin States as a Function of Slope Angle, 1971   50
30. Sulfur Content of Estimated Coal Reserves in Four Appalachian Coal Basin States   50
31. Recoverable Surface Mining Reserves by Estimated Sulfur Content in Four Appalachian Coal Basin States, 1974   51
32. Estimated Life of 1974 Coal Reserves by Sulfur Content Based upon 1971 Production Rates for Four Appalachian Coal Basin States   52
33. Coal Sold Commercially and by Captive Mines in Four Appalachian Coal Basin States, 1969-1973   53
34. The Largest Eastern Kentucky Coal-Mining Firms, 1973 Production   54
35. Thirty-three Largest Coal Mines in Eastern Kentucky, 1973   56-57
36. Eastern Kentucky Coal Production, 1974   58-59
37. Growth in Coal Output by Eastern Kentucky Counties, 1969-1974   62

38. Size Distribution of Mines among Four Appalachian Coal Basin States, 1973   63
39. Average Mine Output for Four Appalachian Coal Basin States, 1973   64
40. Percentage Distribution of Mines and Output by Size for Four Appalachian Coal Basin States, 1973   65
41. Percentage Share of Output from Different Size Underground Mines in Four Appalachian Coal Basin States, 1969-1973   70
42. Percentage Share of Output from Different Size Surface Mines in Four Appalachian Coal Basin States, 1969-1973   71
43. Surface Mines Licensed in Eastern Kentucky, 1970-1974   73
44. Productivity in Coal Mining for Four Appalachian Coal Basin States, 1968-1973   74
45. Productivity in Coal Mining for Four Appalachian Coal Basin States, 1968 and 1973   75
46. Eastern Kentucky Coal-Producing Counties in Descending Order of Strip-Mining Productivity, 1973   79
47. Eastern Kentucky Coal-Producing Counties in Descending Order of Underground-Mining Productivity, 1973   80
48. Eastern Kentucky Coal-Producing Counties in Descending Order of Auger-Mining Productivity, 1973   81
49. Number of Days Worked by Coal Mines in Four Appalachian Coal Basin States, 1969-1973   84
50. Unionization in Coal Mining, 1968   84
51. Active Mining Days as a Function of Mine Size in Eastern Kentucky, 1972   85
52. The Price of Coal, 1968-1973   87
53. Percentage of Coal Sold Commercially and by Captive Mines in Four Appalachian Coal Basin States, 1969-1973   89
54. Net Incomes of the Major Coal-Producing Companies in the U.S., 1973-1975   92-93
55. Analysis of Tipple and Delivered Samples in Western Kentucky, Indiana, and Illinois, 1973   97
56. Coal Production in the U.S. and Western Kentucky, 1967-1975   97
57. Share of Western Kentucky, Indiana, and Illinois in National Output, 1967-1975   98

58. Total Value of Coal Mined in the U.S., Western Kentucky, Illinois, and Indiana, 1967-1974   99
59. Comparative Values of U.S., Western Kentucky, Indiana, and Illinois Coal, 1967-1974   100
60. Average Number of Men Working Daily in Coal Mines in Western Kentucky, Illinois, and Indiana, 1967-1973   101
61. Total Primary and Secondary Employment in the Eastern Interior Coal Basin in 1973   102
62. Total Recoverable Reserves in Short Tons and Percentage of the Western Kentucky Coal District Recoverable Reserves per County   103
63. Western Kentucky Coal District Coal Reserves and Life beyond 1971 at 1971 Rates of Production   104
64. Estimated Remaining Coal Reserves of the Eastern Interior Coal Basin, January 1, 1974   105
65. Origins of Demand for Coal Mined in the U.S. and Eastern Interior Coal Basin, 1973   106
66. Categories of Electric Power Use in the U.S., 1965-1990   107
67. Share of Electric Utilities in Total Coal Consumed in the U.S. and Eastern Interior Coal Basin, 1967-1973   108
68. Comparison of Forecasts on U.S. Electricity Consumption   108
69. Origin of Demand for Coal Mined in the Eastern Interior Coal Basin, 1969-1973   109
70. Market Shares in Output of the Eastern Interior Coal Basin, 1973   112
71. Major Markets for Indiana and Illinois Coal, 1973   112
72. Market Shares of Illinois and Eastern Interior Coal Basin   113
73. Markets for Western Kentucky Coal, 1969-1973   116
74. Rank-Size Distribution of Markets for Western Kentucky Coal, 1969-1973   118
75. Production of Coal, Eastern Interior Coal Basin, 1974   119
76. Distribution of Coal Production among U.S. Districts, Eastern Interior Coal Basin, 1967-1974   120
77. Years of Theoretical Life of Coal Reserves, Eastern Interior Coal Basin   121
78. Theoretical Life of Strippable Coal Reserves for Selected Counties in Western Kentucky   121
79. The Largest Mines in the Eastern Interior Coal Basin, 1973   122

80. Coal Output by Parent Firms in the Eastern Interior Coal Basin, 1973   123
81. Average Size and Productivity of Labor for Mines in Indiana, Illinois, and Western Kentucky, 1970 Production Year   126-27
82. Size Distribution of Mines among the Eastern Interior Coal Basin States, 1973   128
83. Percentage Distribution of Mines by Size of Output, Eastern Interior Coal Basin, 1973   128
84. Summary of an Engineering Study of the Cost of Mining Coal in Western Kentucky   130
85. Percentage Share of Output from Different Size Categories of Underground Mines, 1968-1973   131
86. Percentage Share of Output from Different Size Categories of Surface Mines, 1968-1973   132
87. Percentage Distribution by Type of Mine, 1967-1974   134
88. Average Railway Transportation Costs per Ton of Coal, U.S., 1967-1973   135
89. Representative Rail Charges for Coal, May 1968   136
90. Cost of Transporting Coal by Unit Train, 1969   137
91. Methods of Transporting Indiana, Illinois, and Western Kentucky Mined Coal, 1973   138
92. Productivity and Price in Coal Mining, Eastern Interior Coal Basin, 1967-1973   140
93. Per Capita Personal Income, 1968 and 1973   144
94. State Implementation Plan Sulfur Content Limitations from Regulations and Conversions, as Used in Supply/Demand Calculations   145-47
95. Recoverable Energy Resources by Type of Fuel and World Region, 1970   152
96. Estimated Recoverable Reserves, U.S.   155
97. Estimates of Supply of Continuous (Renewable) Energy (10^{12} Watts) for the U.S.   156

Maps

1. Coal Deposits in Kentucky   15
2. Distribution of Coal in the Eastern United States   18
3. Major Coalfields of the United States   19
4. Regional Distribution of Eastern Kentucky Coal Shipments to Electric Utilities   34
5. Kentucky Coal Output by County, 1974   61
6. Geographic Market Areas for Imported Residual Fuel Oil   91
7. Eastern Interior Coal Basin   96
8. Regional Market Distribution of Western Kentucky Coal, 1973   114

Figures

1. The Energy/GNP Ratio, 1947-1970   25
2. Relationship between the Percentage of Underground-Mining Operations and the Percentage of Coal Production, 1973   66
3. Relationship between the Percentage of Surface-Mining Operations and the Percentage of Coal Produced, 1973   69
4. Productivity in Coal Mining for Four Appalachian Coal Basin States, 1968-1973   77
5. Relationship between Productivity and Mine Size in Underground and Auger Mining in Eastern Kentucky, 1973   82
6. Relationship between Productivity and Strip Mine Size in Eastern Kentucky, 1973   83
7. Price of Coal in Selected Appalachian Basin States, 1968-1973   88
8. Demand for Coal by Electric Utilities   110
9. The Rate of Growth in U.S. Energy Consumption and Population, 1800-2000   111
10. Labor Productivity in Mining for Indiana, Illinois, and Western Kentucky by Counties, 1970   124
11. Relationship between Price and Productivity for the Eastern Interior Coal Basin by State, 1967-1973   141
12. Manufacturing Job Increases in the United States and the Southeastern Region, 1950-1970   143
13. State Sulfur Regulations for Coal   148
14. Cumulative Coal Availability/Requirement at SIP Regulations, 1975   149
15. Coal Availability/Requirements at SIP Regulations, 1975   150

Preface

This book presents the results obtained from examining the economics of the Kentucky coal industry. The analysis inquires into the demand-and-supply properties of Kentucky coal and attempts to discover the nature of the economic framework within which the industry functions. The study does not represent a particular position, nor does it develop a unique point of view. Rather, it attempts to integrate and interpret those economic and institutional variables that influence the competitiveness and welfare of the Kentucky coal industry.

To an extent, the scope of the book is circumscribed by the unavailability of sufficiently detailed data. It thus became necessary from time to time to draw inferences and conclusions from other than strictly empirical data sources. By and large, however, a sufficiently comprehensive picture emerges to broaden our understanding of the economics of this important industry.

The audience that this book addresses is varied. Researchers in the field of energy, and in particular in coal as an energy input, will find useful information on the role of the industry as an energy supplier. Likewise, policymakers whose responsibilities lie in the energy area will gain insights into the structure of the Kentucky coal industry and, by proxy, into the makeup of neighboring industries. This book is also directed to businessmen, engineers, and brokers whose activities are centered in the energy field.

I wish to thank some of my colleagues and graduate students for their valuable assistance during the preparation of the book—in particular Dr. Stuart Schweitzer and Dr. Phillip Karst as well as Robert Spoerl and J. B. Marshall.

The University of Kentucky Institute for Mining and Minerals Research under the directorship of Dr. James Funk provided considerable support for the work described in this book and its generosity is gratefully acknowledged. The responsibility for the results and conclusions, however, rests entirely with me.

1. The Demand and Supply of Kentucky Coal

To come to grips with what is commonly regarded as the energy problem, it is important to consider briefly the composition of our energy-resource inventory which consists of oil, natural gas, nuclear power, and coal. The first three—the glamor resources of the post-World War II era—have been the subject of voluminous research studies for years; coal has not. Its appeal to serious researchers has been minimal at best, probably because coal has been a declining industry until only very recently.[1] Following World War II, the railroads, then coal's largest customer, switched from coal-fired steam locomotives to diesel engines, and other users likewise shifted to cleaner and more convenient fuels—principally natural gas and oil. Relative price levels were such as to make this shift economically attractive.

By the end of the 1960s, the coal industry had begun to recover from the blow of a 36 percent decline in output, and today coal issues are once again beginning to capture the interest of industry analysts. This interest was bolstered significantly by the recent events in the world petroleum market. The rapid emergence and success of the Organization of Petroleum Exporting Countries have served notice that major changes in our energy-resource inventory and in the use we make of it must be expected. But the precise nature of these changes and of the policies needed to implement them remains a matter of deep controversy.

This book, in focusing on the economic structure of the Kentucky coal industry, attempts to fill a research void which has existed for many years. No comprehensive analysis of this Kentucky industry has ever been undertaken, even though the state continued unabated to maintain its more than one-fifth share in national coal output, reaching 140 million tons, or 22 percent of national production in 1975 (see Table 1).

Before embarking on an analysis of the Kentucky coal industry and its component parts, I will consider briefly the analytical framework within which this study is set. This framework is structured around the traditional concepts of demand and supply and shows how each influences the volume and type of coal that flows to markets.

Table 1. U.S. and Kentucky Coal Production, 1960-1975

	U.S. production (million tons)	Kentucky production (million tons)	Percentage of U.S. production
1960	416	68	16
1965	512	78	15
1970	603	125	21
1971	552	120	22
1972	595	121	20
1973	592	128	22
1974	603	137	23
1975	640	140	22

Sources: Bureau of Mines, *Minerals Yearbook* and *Mineral Industry Surveys,* 1961, 1966, 1971-1975.

THE DEMAND FOR COAL

The demand for coal, like the demand for any natural resource, is derived essentially from the demand for the final commodity in whose production it participates as one of many inputs. Very little demand for coal originates today with consumers who intend to use it for home heating. The direct-heating segment of coal demand has passed into history. The demand for coal for use in the production of electricity, on the other hand, has risen sharply in the period after World War II and in 1973 absorbed nearly two-thirds of national output.

Product-demand relationships are frequently summarized in mathematical shorthand showing in functional form the variables on which the quantity demanded of a commodity depends. For example, a demand function for coal might be written generally as:

$$D_c = f(P_c; P_1, P_2, \ldots P_n; Q_1, Q_2, \ldots Q_n; K_c, K_1, K_2, \ldots K_m).$$

This equation tells us that the amount of coal demanded (D_c) at any time is a function of the variables enumerated in the parentheses. P_c denotes the price of coal and is the only variable whose value depends directly on the equation because the price of coal is set by the interaction of supply and demand forces in the coal market. Variables $P_1, P_2, \ldots P_n$ represent the prices of fuels substitutable for coal in its several uses. As a rule, more coal will be demanded when the price of substitutes, such as fuel oil, rises. Similarly, less will be desired when the prices of fuel substitutes fall. Variables $Q_1, Q_2, \ldots Q_n$ show the amount produced of electricity, steel, or other products which use coal in their production. The final set of variables, $K_c, K_1, K_2, \ldots K_m$ reflects more-or-less constant qualitative properties of coal and of its substitutes (for example, energy or sulfur content), which are relevant to demand decisions for coal.

The key variable that governs the demand for any product is the product's price, with demand inversely related to price. This means that, setting aside the influences of other factors, the lower the price of any commodity, for example, coal, the more of it will be demanded. Likewise, the higher its price, the less of it will be demanded. The exact price-quantity relationship, that is, the sensitivity of demand for a commodity to changes in its price, is subsumed under the principle of the price elasticity of demand and can generally be measured empirically. But a paucity of data and the constraint of resources represent formidable estimation obstacles and have rendered efforts designed to identify price-elasticity coefficients beyond the scope of this study. It is sufficient to note that a producer is best off if he faces an "inelastic" demand for his product (that is, if he faces a noncompetitive market in which he can exercise influence) because, by definition, he can raise his price and still be assured that, even though the demand will decline somewhat, total revenue will rise. In the case of an "elastic" demand for his product, on the other hand, the producer would be loath to raise prices because the resultant reduction in quantity would be so large as to lead to a decline in total revenue.

Four basic factors influence the price elasticity or inelasticity of demand for any product, including coal. The most important factor is the availability or unavailability of substitutes. The greater the number of usable substitutes for coal, the greater will be the price elasticity of demand for it, or, put differently, the greater will be the adjustment in quantity demanded to changes in price. Recent increases in coal prices have had little impact so far on the quantity of coal demanded for use in electric-power generation, because oil, gas, and nuclear prices have risen sharply and the threat of an oil embargo is ever present. Similarly, there exists no economic substitute for coking coal—coal used in the production of pig iron. As a consequence, the quantity demanded of this type of coal is also insensitive to changes in its price.

A second factor that influences the elasticity of demand for a commodity is the fraction of total production costs that is accounted for by the item being examined. The smaller this fraction, the less elastic is the demand for the product as an input because changes in the input price would exert little influence on total production costs. Relatively low coal prices in the past have contributed to the relative inelasticity in the demand for coal. In the past, fuel costs have accounted for no more than 2 to 3 percent of the manufacturing costs.[2] But the recent rise in the price of fuels, including coal, has raised this proportion substantially so that it is reasonable to conclude that the demand for coal is less insensitive to prices today than it was only a short time ago and that dynamic elements have become very important.

In conjunction with this factor is the third. The individual price elasticities for coal also depend in part on the elasticities of the final commodities or services in whose production coal participates. The price elasticities of demand for electric power or steel, for instance, do influence the price elasticity of demand for coal.

The fourth factor that has an observable impact on the price elasticity of demand for any product is the passage of time. The longer the time period over which a price-supply relationship is examined, the greater the observed changes in quantity demanded in response to changes in price. The longer the period, the more time buyers have to make adjustments in input combinations. For example, a decline in the relative price of coal will, over some period, enable potential buyers of coal to obtain new equipment and develop new production processes which capitalize on the relative attractiveness of coal at its lower price.

In addition to price, the various characteristics and properties of commodities also have an influence on the quantity demanded. In the case of coal, chemical composition and geographic location are now two of the most influential features.

The most basic of the relevant chemical properties of coal is its heat content measured in British thermal units (BTUs) per pound. Coal users buy energy, and, other things equal, the more of it per dollar expenditure they can obtain, the lower will be their unit costs of production; hence the more of it they will desire. In addition, the amount of sulfur and ash contained in the coal also influences the demand for it. Air-purity standards in certain locations prohibit the burning of coal with high sulfur and ash content; in others, prohibitions are less stringent.

THE SUPPLY OF COAL

The amount of coal that firms desire to sell at any one time for various prices is called the supply of coal. Like the supply of any commodity or service offered for sale, the amount actually entering the marketplace depends upon a wide range of factors. Some of these are easily identifiable, others are not. If, for example, a high degree of risk attends the future salability of a particular commodity, producers may shy away from its production unless compensation for risk is satisfactorily reflected in the price they expect to receive. It is important, however, to distinguish between quantity offered for sale and quantity actually sold. The first need not be coterminous with the second, the latter being a reflection of what a firm actually succeeds in selling, which may or may not be equal to that which it desires to sell.

The willingness of a firm to sell certain products or certain quantities of a particular product—for example, coal—is influenced by a variety of factors. For purposes of convenience these factors may be classified into four categories. First, the quantity of coal supplied depends upon the goals of the coal-mine operator. If profit maximization is the primary goal of the firm, then an amount equal to the quantity at which the firm maximizes profits, given the expected price, would be offered for sale. If, on the other hand, other objectives dominate, operators may offer for sale a quantity of coal that is inconsistent with that which would have been offered under the profit-maximization goal. An example of a nonprofit maximization objective might be the desire of a firm to continue operations and provide employment for a local community for social reasons, even though market prices are temporarily low and would not cover all production costs.

Second, the quantity of coal supplied depends upon the state of mining technology. New mining techniques and the focus on surface mining have greatly increased the ability and willingness of mine operators to offer for sale, at any given price, various quantities of coal. For example, while an operator using outdated mining techniques might have been unwilling to offer any coal for sale at $6.00 per ton, with the development and adoption of new production techniques he may be willing and able to offer for sale large quantities at that price.

Third, the quantity of coal supplied depends upon the price of coal and the prices of coal substitutes and other commodities. It is quite clear that the higher the price of coal that the mining firm can obtain for its output, the greater will be its incentive to produce. The typical firm is assumed to be guided by the profit motive. In part also, the amount of coal offered for sale depends upon the prices of all other commodities and in particular upon prices of related commodities such as natural gas and fuel oil. If, for example, relative prices change and profits in other industries are greater than in the coal industry, it is quite conceivable that fewer new investments will be made in the now relatively less profitable coal industry and more in the other more profitable industries. The amount of coal offered for sale, therefore, will be less than it would have been had this change in relative prices and profits not occurred.

Finally, the quantity of coal supplied depends upon the costs of mining coal. If the price of one or more of the factors of production that go into the mining of coal increases, then for any given market price of coal, less will be offered for sale than if factor prices had remained unchanged. The reason for this can be traced to the impact of higher factor prices on the profitability of the firm if it is unable to pass all the higher factor costs to the buyers of coal in the form of higher prices. In a purely competitive environment, the individual coal-mine operator would be unable to increase the price at which he sells his coal without risking loss of sales and may shut down operations.

Increased union wage payments, higher mineral royalties, and rising costs of mining equipment are all examples which could lead a noncompetitive coal-mine operator to reduce output in an effort to pass on these higher costs to the purchasers of coal, unless, of course, he can cut other costs such as nonunion wages or social costs.

Product-supply relationships are frequently summarized in mathematical shorthand showing in functional form the variables on which the quantity of a commodity supplied depends. For example, a supply function for coal might be written generally as:

$$S_c = f(G, T, P_c, P_1, \ldots P_n, r_1, r_2, \ldots r_m),$$

where G represents the goals of the coal-mining firm, T the state of mining technology, P_c the price per ton of coal, $P_1, \ldots P_n$ the prices of coal substitutes and all other commodities that influence the supply of coal, and $r_1, r_2, \ldots r_m$ capital costs, mineral rights payment, etc. The equation tells us that the amount of coal which firms are willing to sell at any one time, S_c, is a function of the variables enumerated in the parentheses.

The key variable that governs the supply of coal is its price, with supply positively related to it. Disregarding the influence of other factors, the higher the price of coal, the greater will be the desire of coal operators to supply it. The lower the price, the weaker will be this desire.

The exact price-supply relationship (that is, how sensitive the quantity of coal that firms are willing to supply is to changes in the price of coal) is subsumed under the principle of the price elasticity of supply. To measure this elasticity empirically, however, is a most complex task. Certainly if price changes elicit no response, the supply appears totally inelastic; and a large response would indicate a price-elastic supply.

Many of the factors that influence the price elasticity of demand also have an influence upon supply. In general, if mining costs increased very rapidly with the expansion of mining operations, firms would have a weak incentive to expand output in the event of a rise in the price of coal. In this case, supply would be relatively price-inelastic. If, on the other hand, mining costs increased only slightly with rising output, then the firms would have a strong incentive to raise output as prices rose, and supply would be price-elastic. This incentive is derived from the assumed desire of each mine operator to make as large a profit as possible.

To relate time to changes in output, one would say that the longer the period under analysis, the more elastic supply becomes. Given an ample amount of time, nearly all supply will adjust to changes in price. In the short run, however, too little time exists for firms to adjust their output to price changes, and supply therefore is said to be price-inelastic.

Before we proceed to an analysis of the eastern and western Kentucky coal industries, it is useful to examine briefly the characteristics of the United States coal industry within which the Kentucky regions operate. In 1975 United States coal mines produced 640 million short tons of coal, an output 6 percent higher than in 1974. Had the United Mine Workers not gone

Table 2. Annual Production of the U.S. Bituminous Coal and Lignite Industry, 1969-1975

Year	Production Millions of Short Tons
1969	560.5
1970	602.9
1971	552.2
1972	595.4
1973	591.7
1974	601.0
1975	640.0

Sources: Bureau of Mines, "Coal—Bituminous and Lignite," *Mineral Industry Surveys,* 1970-1975.

on strike in late 1974, production in that year would have exceeded the 603 million-ton level, and output in 1975 would have been correspondingly lower. The work stoppage that commenced in the second week of November led to such a sharp reduction in output that by the fourth week of the month, a mere four million tons per week were being mined. Table 2 shows the annual levels of U.S. production for the past seven years. For the last six years production levels have remained relatively stable (1975 output is only 6 percent higher than 1970 production), particularly when one considers the impact which recent shortages and price increases in fuel oil might have had on the industry. Even if the mine workers' strike had not taken place, output in 1974 probably would not have increased more than 6 to 7 percent over 1973.

Coal-mining firms, like firms in most other industries, adjust their output and prices to the type of market conditions they face. The competitive structure of the industry or of particular subregions therefore influences strongly the behavior of individual coal-mining firms. Whether prices and output are determined competitively can be inferred by examining the structure of the industry from three vantage points.

One of the traditional measures used in attempts to establish the competitiveness of an industry is to determine how much of total output is provided by how many firms. In the coal industry, a large portion of output is provided by two coal producers: the Peabody group and the Consolidation group (see Table 3). Together, these two mining groups accounted for 22 percent of national output in 1973. The fifteen largest coal companies produced slightly more than one-half the total output.

There exists no unanimity among experts on whether the concentration ratios shown in Table 3 represent a significant degree of market concentration. Judgments vary from moderate to no concentration. For example, two

Table 3. Largest Commercial Bituminous Coal Producers in the U.S. during 1973
(ownership, date of acquisition, and percentage of total commercial production)

Rank	Coal Firm	Ownership or Controlling Company	Date of Acquisition	1973 Commercial Production (Million Tons)	Percent of Total Commercial Production[a]	Cumulative Percent
1	Peabody Group	Kennecott Copper	3-29-68	69.9	11.83	11.83
2	Consolidation Group	Continental Oil	9-15-66	60.5	10.23	22.06
3	Island Creek Group	Occidental Petroleum	1-29-68[b]	22.9	3.87	25.93
4	Pittston Group	Pittston Co.		18.8	3.18	29.11
5	Amax Group (Ayshire)	American Metal Climax	10-31-69	16.7	2.82	31.93
6	U.S. Steel (captive)	U.S. Steel		16.2	2.74	34.67
7	Bethlehem Mines (captive)	Bethlehem Steel		14.1	2.39	39.06
8	North American Coal Corp.	North American Coal Corp.		12.5	2.12	41.18
9	Old Ben	Standard Oil of Ohio	1968	10.8	1.84	43.02
10	Eastern Associated	Eastern Gas and Fuel Associates		10.6	1.80	44.82
11	Westmoreland	Westmoreland Coal Co.		8.8	1.49	46.31
12	General Dynamics Group	General Dynamics		8.7	1.47	47.78
13	Pittsburgh and Midway	Gulf Oil	Late 1963	8.1	1.36	49.14
14	Utah International	Utah International		7.4	1.25	50.39
15	American Electric Power	American Electric Power Service Corp.		6.6	1.11	51.50

[a]Total commercial production during 1973 was 591 million tons.

[b]Acquisition of Maust Coal Properties, July 8, 1969

Sources: *Keystone Coal Industry Manual*, 1973, 1974.

Table 4. Concentration Ratios for Total Bituminous Coal Production in the U.S. for the Years 1950, 1960, 1970, and 1974[a]

	Percent of Total Production[b]			
	1950	1960	1970	1974
Single Largest	4.8	7.0	11.4	11.3
2 Largest	9.1	13.9	22.1	19.9
3 Largest	11.6	18.2	27.1	23.4
4 Largest	13.6	21.3	30.5	26.7
8 Largest	19.4	30.4	41.0	44.3
12 Largest	23.6	36.5	48.9	50.4
15 Largest	26.4	39.6	52.2	54.1
20 Largest	30.4	44.4	56.5	58.5
50 Largest	45.2	59.9	68.3	72.8
Producers of at least 100,000 Tons	82.8	87.0	93.8	94.5
Remaining	17.2	13.0	6.2	5.5

[a]Includes both commercial and captive production. A few enterprises listed as firms might be technically termed as affiliated production groups.

[b]Total bituminous coal production in the United States was 516,311,053 tons during 1950; 415,512,347 tons during 1960; 602,932,000 tons during 1970; and 603,000,000 tons during 1974.

Sources: Calculated from *Keystone Coal Industry Manual*, 1950, 1960, 1970, 1974; Phillip E. Giffin, *Industrial Concentration and Firm Diversification in Bituminous Coal with Special Reference to the Southeastern United States, 1950-1970* (Knoxville: University of Tennessee, 1972).

authors maintain that if eight firms sell 50 percent and twenty firms, 75 percent of national output, an industry may be considered highly concentrated.[3] Clearly this is not the case in the coal industry. Another author, S. N. Whitney, states, "There is a relatively low degree of concentration in bituminous coal compared with most other industries."[4]

The concentration issue is further clouded by the existence of captive output, that is, output that never enters the market because it is produced for internal use only. Steel companies that use coal produced by their own mines are an example of this. More importantly, large quantities of coal are sold on long-term contracts over periods of ten or more years. This coal also does not enter the market directly, but rather, influences market prices indirectly and only in the long run. We can note, therefore, that nationally, the coal industry can be characterized as moderately concentrated at best.

An important question to be asked is whether significant changes in market concentration in the coal industry have occurred during the post-World War II era. The evidence suggests that up to approximately 1970, concentration in the industry did increase (see Table 4). For example, by 1970, the two largest firms were producing 22.1 percent of national output while twenty years earlier they accounted for only 9.1 percent. For the twenty largest firms the ratio is 56.5 to 30.4 for the same period. And for all coal producers with an output of more than 100,000 tons per year, the collective market share rose from 82.8 to 93.8 percent. The increase in concentration for the twenty years preceding 1970 took place mostly in the twenty largest firms, while the smaller firms kept their market shares intact. Between 1970 and 1974, there occurred a moderate reduction in concentration in the industry, but this change is neither large enough nor sustained over a sufficiently long period of time to represent a trend. Its significance is best interpreted as reflecting the sizable influx of relatively small firms into the industry, largely in surface mining and in response to the rising price of coal.

The static theories of competition emphasize that the single most important element that fosters concentration in industries is a barrier to the entry of new firms.[5] Although there is some disagreement among scholars on this point and on the exact definition of entry, it is fair to say that no monopoly can endure unless new firms are prevented from entering the industry. If monopoly power does exist in any industry, it typically results in higher than competitive prices, higher production costs, and lower output with fewer resources employed.

Key determinants of ease of entry are resource mobility and divisibility and knowledge. If initial investment costs are low and factor resources readily available and divisible, entry would be relatively easy. Restrictions on any of these conditions would obstruct entry.

When considering the issue of entry in the coal industry, a distinction must be drawn between underground and surface mining. In the former, the size of the initial investment is large, with an estimated $20 to $30 million required today for a mine expected to produce approximately one million tons per year. Variations in these initial capital requirements can be significant, depending on geological structures and formation, soil stability and composition, and seam depth.

In response to the requirements set forth in the Federal Coal Mine Health and Safety Act, initial investment costs have risen substantially as more elaborate and well-defined safety equipment must now be installed in new mines. Thus the entry costs into underground mining are getting higher and in the 1970s entries fell due partly to the more stringent safety requirements and partly to the general escalation of capital-equipment costs. It is no longer a simple matter to open an underground mine, and the days of the small, family-type operation may be numbered. Entry into the industry is confined to firms with access to considerable investment capital.

Although surface-mining operations also require substantial capital equipment, the initial investment cost is significantly smaller than for underground operations. In fact, if the capital equipment is leased, initial investments can be kept rather small. More importantly, much of the initial investment is more divisible, since surface operations can be started on a small scale, to be expanded at a later date.

A large portion of the equipment used in surface mining—bulldozers, trucks, loaders, graders—is also used in the heavy-construction industry for the building of roads, bridges, and commercial facilities. As a consequence, this type of equipment is substitutable between the two industries. When construction activity slows, this equipment becomes available for surface mining. And since this is rolling as opposed to stationary equipment, it is easily transferred from one region or industry to another.

The evidence suggests that entry into underground mining is restricted by a host of economic, institutional, and technical factors, while entry into surface mining is not. It is therefore not surprising that short-run supply is much more sensitive to price in surface mining than in underground mining. A surface mine can be opened or expanded in a matter of months; an underground mine requires years. Likewise, however, surface operations can be abandoned quickly, while underground mines tend to operate as long as their variable costs and at least a portion of their fixed costs are met. Exit from underground mining is more costly than from surface mining; this fact leads to a somewhat greater stability in output of the industry. It is correct to conclude, therefore, that the potential for competitive behavior, based upon the criterion of ease of entry, is considerably greater in surface than in underground mining. A decline in the share that surface mining holds in total output would therefore contribute to a reduction of the competitive potential in the industry.

Coal is used primarily for the production of steel, which uses coking or metallurgical grade coals, and the producing of steam, which uses steam coal. These two usages clearly delineate the markets for coal and there has generally been little substitution between the two types of coal in the past, although this may not be so in the future.

Within each of the two submarkets, considerable product homogeneity exists in regions. Environmental regulations, and particularly air standards, if they become more uniform and universal, will tend to strengthen product homogeneity in the steam-coal market. This means that it is unlikely that individual coal producers will be able to distinguish their products in the eyes of buyers as much in the future as they have in the past. Coking coal has to satisfy certain technical parameters, and it matters little to the buyer where or by whom the coal is mined, as long as the purchase costs are minimized. Similarly, it matters not where steam coal originates as long as BTU, sulfur, and ash levels meet the necessary environmental and cost standards. No one producer is likely to be able to greatly distinguish his coal from competing producers, assuming the cost is equal.

A majority of the largest coal firms are subsidiaries of companies whose main activity lies outside the coal industry. Table 3 shows that, in 1973, only two of the fifteen largest coal producers in the United States were independent operators active in the coal industry alone. All other producers were subsidiaries of companies whose main interests lay outside the industry. Together, these two companies, the North American Coal Company and the Westmoreland Company, produced only 3.61 percent of the national output. The remaining thirteen companies were distributed over several industries—oil, electric power, steel, metals, and conglomerate.

The incentive for steel companies to produce coal is straightforward, "no substitutes for coal exist in the iron-making process."[6] Therefore, a guaranteed source of this critical input would seem to be a cardinal objective of steel producers even if the costs are moderately higher than if the inputs were purchased competitively. Since coking coal produced for internal use does not enter the marketplace, it has no direct impact on price, and its supply is not influenced by changes in coal prices. Market price becomes merely an accounting price to the steel companies. Excess output, that is, coal not consumed internally, could enter the market, but the amount would not be very great. Thus only a portion of the total coking coal mined each year enters the marketplace; a large share is absorbed directly by the parent steel firms.

By far, the largest consumer of coal in the nation is the electric utility industry, and it is here that, in the long run, coal competes with other fuels. The substitution of residual fuel oil and natural gas for coal, or vice versa, is feasible and, depending upon relative prices, can be attractive. In fact, some utilities have the technical capacity to generate electric power with either oil or coal. But substituting coal for fuel oil or gas is more difficult and considerably more expensive.[7]

In order to reduce the risk of not being able to supply coal-generated electric power because of interruptions in the supply of coal, most utilities stockpile coal supplies for as many as ninety days and purchase it on long-term contracts. Some of these contracts are written for periods exceeding a decade. The steel companies likewise attempt to reduce risks by the outright purchase of coal-mining operations called captive mines. Either behavior, although perfectly rational from the standpoint of risk aversion, tends to reduce either the number of active firms or the amount of output entering the market, and thereby the intensity of price competition. It has been estimated that as much as 80 percent of coal output is sold on long-term contracts, coal that is insensitive to short-run price fluctuations. If this output is added to the output from captive mines, then it is easy to understand why short-term fluctuations in demand frequently have a rather profound impact upon price, because little coal, in relative terms, remains for sale on the open market. Contract prices on the other hand tend to remain relatively stable and adjust only with time.[8]

In sum, it is possible to distinguish between two basic coal markets: coking and steam. In each of these, two types of market forms are operative: the spot market (short term) and the contract market (long term).

Statements abound from critics of the coal industry alleging conspiratorial behavior of large coal producers with intent to restrict output and raise prices.[9] Evidence to substantiate these allegations, however, is at best inconclusive. It is, of course, correct that during 1974, the price of coal tripled or quadrupled in many coal markets and that this increase was far in excess of cost increases. However, it is also quite clear that the main force behind the price rise was not cost escalation but unusually strong demand. In 1974, buyers were actively stockpiling coal in anticipation of a labor strike, many public utilities began switching from oil to coal-fired operations, steelmakers continued to manufacture steel and demand coking coal despite obvious slowdowns in the economy, and railcar bottlenecks were numerous with uncertain delivery dates. Thus, while the possibility of restrictive market behavior on the part of some coal producers must not be discounted, the evidence does not support the conclusion that such restrictive behavior was the principal cause of the rapidly escalating prices of coal. And in 1975 coal prices dropped precipitously.

The past twenty-five years have seen some increase in producer concentration in the United States industry, but this increase has not progressed far enough to make collusive activity real, and poor economic performance is its consequence. What is more, during the early 1970s, this greater concentration seems to have abated (see Table 4). In some regional submarkets concentration may be uncomfortably great, as, for example, in Illinois, "where seller concentration is high and the courts have already called for a divesture in a coal merger."[10] In other submarkets, however, relative concentration is low and competition vigorous. Thus, when addressing the issue of competition in the coal industry, it is best to delineate a particular regional submarket and examine it rather than the industry as a whole. Kentucky's coal markets are generally bounded by transportation cost considerations, just as Rocky Mountain coal markets are. And although there have been some significant shifts in these boundaries of late, the general delineations do hold. For example, it is quite unlikely that eastern Kentucky coal could compete with Montana coal for a share of the Dakota market, just as Utah coal is unlikely to find its way to Alabama or other eastern markets.

Another aspect of potential reductions in the competitiveness of the coal industry deals with conglomerate or oil-company purchases of coal producers. Whether such integration will in fact lead to restrictive trade and pricing practices is indeterminable. But a recently completed Ford Foundation study concluded that "from a public policy point of view it is hard to justify coal-oil mergers—either those that have occurred or future mergers—if we intend to maintain vigorous competition in the energy sector. The benefits from such mergers have not been demonstrated; the potential injury to competition from a concentration of substitutable energy sources could be substantial."[11]

The Federal Trade Commission recently found that some danger to competition may be engendered by Kennecott Copper's continued ownership of the Peabody Coal Company. But even though the United States Supreme Court upheld the divesture order to Kennecott Copper in May 1974, the issue of whether such integration would have been harmful to competition on economic grounds is far from settled. The only economic rationale that can be used to support the Supreme Court order is that this type of integration tends to lessen competition. But whether a harmful reduction in competition would have resulted from the continued ownership of Peabody by Kennecott cannot be established unless it is first known what the optimum level of competition in the coal industry is and should be.

In conclusion, it is clear that very few large independent mining firms indeed have sprung into operation in recent years. Of the fifty largest coal firms operating in 1973, forty had commenced production since 1950. But of those, only four were of the underground type opened by independent commercial coal companies. The rest were either surface operators, captive mines, or subsidiaries of outside firms.

In Kentucky, coal is one of the most important natural resources, and coal mining an important part of the industrial structure of the state. In the 1970s, the industry increased output steadily, reaching a record annual production of 140 million tons in 1975. In that year, the mining sector of the state consisted of 2,685 licensed mines located in forty-three of Kentucky's 120 counties and provided employment for 41,458 men. Of the coal mined, 75 million tons were obtained from surface operations, 65 million from underground operations. Most coal mines in the state are small; in 1973, only fifty-five mines produced more than 500,000 tons each.

The industry is composed of two distinct coal-producing regions. Each has evolved in response not only to different topographical and geological formations but also in response to the different uses for the two types of coal. Map 1 shows the geographic locations of the coalfields.

Eastern Kentucky, a part of the Appalachian Coal Basin produces high-quality low-sulfur coal, but seams are relatively thin and the topography rugged. Reclamation of surface-mined land is difficult and operations tend to be small. In contrast, western Kentucky, which lies in the southeastern portion

Map 1. Coal Deposits in Kentucky

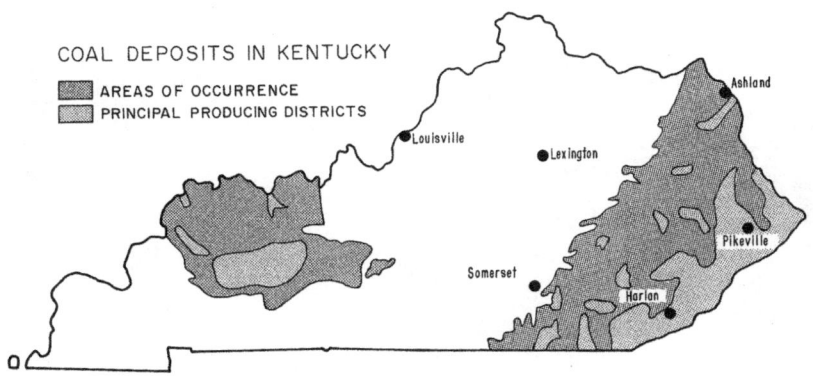

of the Eastern Interior Coal Basin, has thick seams under gently rolling terrain. Land reclamation is much easier in this area. The nature of the topography is also conducive to the development of large-scale mining. This, in turn, raises productivity, which, along with good access to low-cost water transport, puts western Kentucky into a favorable competitive position. Only the high-sulfur content of western Kentucky coal mars what otherwise would be a most optimistic outlook for the industry.

Most of Kentucky coal is sold to four types of buyers: public utility plants, steel companies who combine coking coal with iron to make pig iron, industrial users, and foreign users, who are mostly steel companies. In the past, fluctuations in economic activity have affected most the demand from the last three buyers and least, from electric public utilities.

The energy shortage of 1973-1974 and the dramatic increase in the world-market price of petroleum have had a salutary impact upon the Kentucky coal industry. Demand for Kentucky coal has risen sharply and was met principally from eastern Kentucky mines. In contrast, output from western Kentucky mines declined in 1974, mostly in response to tightened sulfur-emission standards which western Kentucky coal is unable to meet without the aid of sulfur-cleansing equipment.

2. The Eastern Kentucky Coal Industry

The eastern Kentucky coal industry is part of the Appalachian Coal Basin. In addition to eastern Kentucky, the basin includes regions in Pennsylvania, West Virginia, Ohio, Tennessee, Virginia, and Alabama (see Map 2). The last three areas, however, produce relatively little coal; most coal comes from the four major northernmost regions of the basin. The basin clearly dominates the eastern third of the nation and represents an important part of the nation's energy resource base (see Map 3). Table 5 shows the relative quantities of coal produced in 1974 by each of the regions in the basin. In terms of basin output, the four largest regions account for 84 percent of its output, or 48 percent of national production.

Some authorities contend that the coal market spans the entire nation, that the major firms own subsidiaries which "compete with each other throughout the United States."[1] While the exact definition of competition is unclear in this context, it is a fact that coal markets for equal-quality coal are circumscribed by the costs of transporting it. For example, an examination of eastern Kentucky's coal markets shows that with few or insignificant exceptions, sales are confined to markets within an approximate 500-mile radius, and only if coal is shipped by barge, a less costly means of transport, can the market be extended. And even though there exists some overlapping of markets, where coal is shipped depends strongly on transport modes available and their costs.

THE DEMAND FOR EASTERN KENTUCKY COAL

The demand for coal results from the uses to which coal is put. The demand is derived from the end or consumer demand for goods and services in the creation of which coal participates. Eastern Kentucky coal is in strong demand because of its property characteristics.

Although heat content measured in BTUs is perhaps the most important quality-variable that influences the demand for a particular coal, sulfur and ash content can be equally important in certain instances. Eastern Kentucky coal contains relatively generous amounts of energy, approximately 13,406 BTUs per pound, or about 11 percent more than the 12,025 BTUs average

Map 2.   Distribution of Coal in the Eastern United States

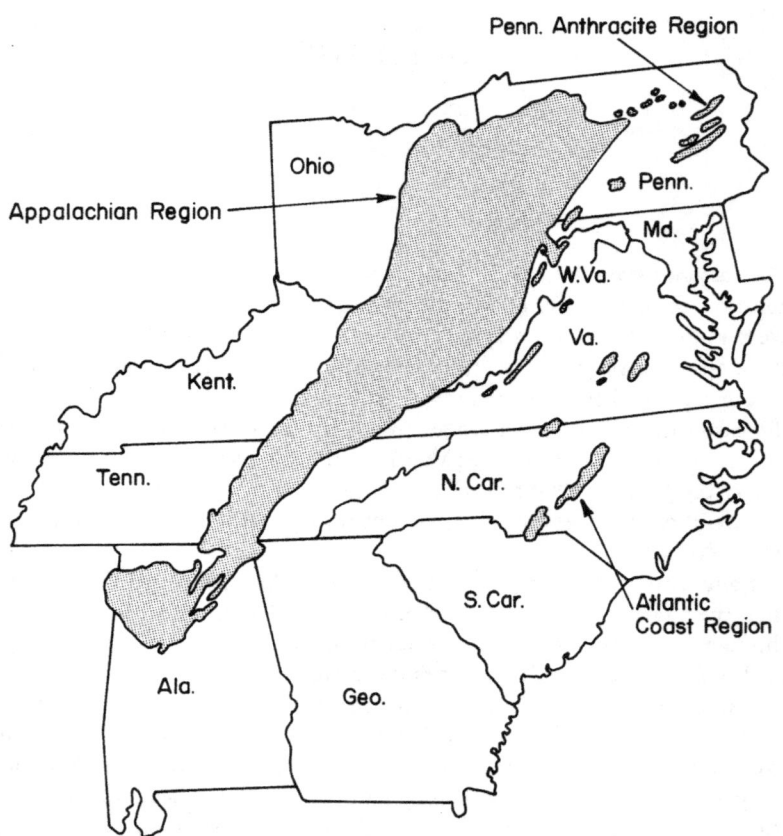

per pound for coal mined in the nation in 1972.[2] Table 6 lists the sulfur, ash, and BTU properties of coal mined in eastern Kentucky, West Virginia, Pennsylvania, and Ohio in 1974 and compares them with the national average for 1972.

The data show that all four areas produce coal that exceeds the national average in BTU content, that the ash content of eastern Kentucky coal ranks only slightly higher than West Virginia coal, and that it ranks lowest in sulfur content.

The ash and sulfur content of coal is important because of the implementation of national air-quality standards and/or the enforcement of local air-quality standards already introduced. These standards limit the amount of ash and sulfur that may be emitted into the atmosphere. In anticipation of

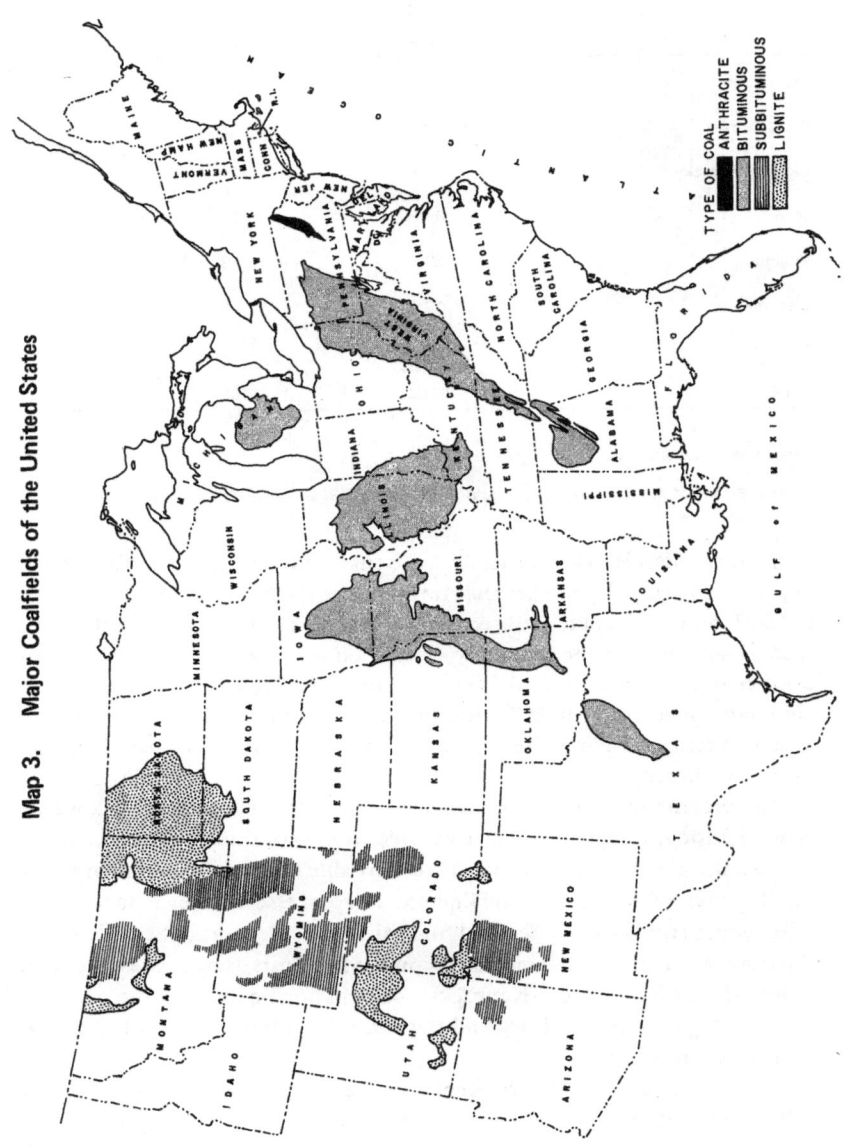

Map 3. Major Coalfields of the United States

Table 5. Production of Bituminous Coal in the Appalachian Basin, 1974

	Output[a] (Million Short Tons)	Percent of Basin Output	Percent of National Output
West Virginia	106.0	29	17
Pennsylvania	78.9	22	12
Eastern Kentucky	77.1	21	12
Ohio	44.6	12	7
Virginia	33.2	9	5
Alabama	19.7	5	3
Tennessee	7.7	2	1
Total	367.2	100	57

[a]Preliminary estimate

Source: Bureau of Mines, "Coal—Bituminous and Lignite in 1974," *Mineral Industry Surveys,* 1975.

these standards, electric utilities began to substitute oil for coal in the process of generating electricity. With the emergence of the Near Eastern oil embargo in 1973, however, and the subsequent upward-spiraling price of oil, the trend toward substituting the relatively cleaner oil moderated, and some electric utilities began to return to coal as their primary fuel source. When oil became abundant once again in 1976 and prices remained stable, electric utilities began to reassess their earlier decisions. Understandably, they are facing an uncertain future.

The present and foreseeable demand for eastern Kentucky coal, however, is quite strong. In part this is attributable to the continued high price of oil. In part also the demand prospects are favorable because future commercial development of gasification and liquefaction processes, if they in fact materialize, could stimulate the demand for coal used in these processes. Also, with the installation of sulfur-removal equipment by plants using coal as an energy input, virtually all eastern Kentucky coal will be in demand, even coal that contains high levels of sulfur. And the demand for low-sulfur coal continues to be very strong.

Nearly one-half of eastern Kentucky's coal reserves are low enough in sulfur content to meet tighter air-purity standards even without sulfur-removal equipment.[3] This ensures that eastern Kentucky's coal market will remain relatively secure and well protected even if more stringent environmental standards are implemented.

Table 6. Sulfur, Ash, and BTU Content of Tipple and
Delivered Samples of Bituminous Coal
(data collected during fiscal year 1974)

Origin of Coal Samples	Ash Content (Proximate %)	Sulfur Content (Ultimate %)	Dry Calorific Value (BTU/pound)
Eastern Kentucky	8.28	0.74	13,406.35
West Virginia	7.92	0.89	13,826.85
Pennsylvania	12.59	1.30	13,029.20
Ohio	12.56	2.69	12,473.77
U.S. average, 1972	N.A.	2.30	12,025.00

Note: All figures are weighted average estimates and no attempt was made to segregate run-of-the-mine from washed coal samples.

Sources: Bureau of Mines, *Analysis of Tipple and Delivered Samples of Coal; Minerals Yearbook,* 1972; U.S. Department of the Interior, News Release, March 10, 1974.

A further property of eastern Kentucky coal is its ability to "coke." Not all coals have coking properties, but eastern Kentucky as well as the rest of the Appalachian Coal region ship more than the national average of their output to coke and gas plants and to overseas users.[4] Virtually all overseas exports of coal are of the coking variety and fully one-half of these originate in Bureau of Mines District 8.

A final characteristic of coal which has a bearing on demand is the geographic location of the coal seam. Because coal is heavy, relative to the number of BTUs it generates, it has a low BTU value-to-weight ratio. This makes coal expensive to transport. In the past, shipping costs represented a large fraction, as much as 60 percent, of the FOB mine price of coal.[5] Producer proximity to users is thus a key element in the competitive structure of the coal market.

We turn now to an analysis of the demand for eastern Kentucky coal which originates with electric utilities, coke plants, other industrial users, retail users, and the export market.

The distribution of bituminous coal by use is shown in Table 7. The statistics point to significant changes in user shares that have occurred in the past twenty-five years. Total output changed little during this period; 1973 tonnage is a mere 10.5 percent greater than 1948 tonnage. But the user distribution of coal, that is, the origin of the demand for coal, has undergone significant alteration. The most vivid change has occurred in the electric utility demand for coal which today absorbs nearly two-thirds of national output.

Table 7. Distribution of U.S. Bituminous Coal Production
by Major User, 1948 and 1973
(million tons)

	1948	Percent of Total	1973	Percent of Total
Electric Power Utilities	95.6	16.9	382.2	64.8
Railroads	94.8	16.7	0.2	0.0
Coke Plants	107.3	18.9	98.0	16.6
Retail Deliveries	86.8	15.3	8.0	1.4
Other Manufacturing and Industrial Users	135.4	23.9	65.8	11.2
Exports	45.9	8.1	35.6	6.0
Total	565.8	99.8	589.8	100.0

Sources: National Coal Association, *Bituminous Coal Facts,* 1972; Bureau of Mines, "Bituminous Coal and Lignite Distribution, Calendar Year 1973," *Mineral Industry Surveys,* 1974.

Railroads no longer use coal; they along with the consumer sector have substituted fuel oil or electricity for coal as their principal energy source. All other uses of coal, in particular manufacturing and industrial uses, have declined significantly over the years, while the export and the metallurgical demands for coal have remained relatively unchanged.

The dominance of the electric utilities' demand for coal is unmistakable on a national plane, and it is, therefore, appropriate to focus first on this category as the principal source of the demand for eastern Kentucky coal.

Electric utilities make up the largest and the fastest-growing segment of the total demand for eastern Kentucky coal. The impact of this segment depends both upon the volume of electric power demanded and on the extent to which coal from eastern Kentucky is used as a fuel to generate electricity.

When measured by any standard, United States consumption of energy in all forms, and of electricity in particular, has grown rapidly. Current consumption of electricity is 300 times that of the 1900s and four times higher than twenty-five years ago. What is more, it is growing at about 7 percent per year.[6]

For most of this century, the growth in electric-power consumption has been greater than the rate of growth of real gross national product (GNP).[7] At the same time the ratio of total energy to real GNP (the energy/GNP ratio) has been declining since the 1920s. The main explanations for this steady

Table 8. The Energy/GNP Ratio, 1947-1970

Year	Amount (Thousands of BTU per 1958 Dollars of GNP)	Percentage Change from Preceding Year
1947	106.1	0.0
48	105.1	−1.0
49	97.5	−7.1
50	96.1	−1.4
51	96.3	+0.2
52	92.6	−3.8
53	91.3	−1.4
54	89.3	−2.2
55	91.2	+2.1
56	94.2	+3.3
57	92.6	−1.7
58	92.8	+0.2
59	91.2	−1.7
60	92.2	+1.0
61	91.7	−0.5
62	89.9	−2.0
63	90.1	+0.2
64	88.7	−1.6
65	87.1	−1.8
66	86.5	−0.7
67	87.2	+0.8
68	88.4	+1.4
69	90.8	+2.7
70	95.6	+5.3

Source: Edison Electric Institute, *Fuels for the Electric Utility Industry, 1971-1985.*

decline are found in the increased thermal efficiency of energy use, the effects of widespread electrification, and the general rise in economic productivity throughout the economy. Table 8 illustrates the behavior of the energy/GNP ratio in the post-World War II era and in column 2 shows annual percentage changes. The decline persisted through 1966, but from that year on, the energy/GNP ratio began to rise. Figure 1 shows the 1966 trend reversal graphically. The exact explanation for this reversal is still under debate, but the major sources have been identified. They are a slowdown or decline in economic productivity; the substitution of electricity for direct fuel use, particularly in residential, commercial, and industrial heating;[8] the increased use of air conditioning; and the increasing use of a proliferating array of electric appliances. These trends in electricity consumption have been exacerbated by continued declines in the relative prices of electricity. When the general price level rises, as it has in the post-World War II period and particularly during the past several years, and electricity prices do not, consumers have a strong incentive to expand their consumption of electricity. This is exactly what has been happening, at least until recently, when higher fuel oil and other prices prompted electric utility companies to begin seeking higher power rates.[9] At present, these rates are rising faster than the general price level. But even so, electricity's cleanliness and versatility, combined with the limited domestic availability of natural gas and, at times, fuel oil for direct use, are likely to prevent a significant decline in the relative importance of electric power.

Published forecasts of the growth in the demand for electricity abound. However, since most of these projections are based upon consumption patterns which either predate the fuel-oil shortage and the price developments of late 1973 and 1974, or are influenced unduly by these events, their predicted values are in question. A possible alternative may be to postulate that the demand for electricity will grow at least as rapidly as GNP adjusted for the effects of fuel conservation and of the continued switching from direct fuel to electric-power consumption. If the two effects were more or less to offset one another, electric-power demand would grow at an average rate of 3 to 4 percent annually, which is the normal real rate of growth of GNP. But if, as is more probable, the switching effect dominates, then demand for electric power will increase more rapidly, and with it the demand for coal.

A useful forecast of United States electric-power generation for the period 1970 to 1985 broken down into ten regions was prepared by National Economic Research Associates. These forecasts show that electric-power generation in the nation is concentrated in six regions, four of which are located east of the Mississippi River, one southwest of the river, including Texas, and the sixth covering the Pacific states and Nevada. While the exact values of electric-power production are less important, Table 9 does highlight the high concentration of power generation and with it the source of future demands for coal as one of the primary fuel inputs.

For the decade of the 1970s, the greatest amount of predicted growth is in the Central and South Central regions of the United States, while beyond 1980, the forecasters envision relatively balanced growth throughout the nation. Only the East Central region seems to be lagging somewhat behind during this latter period.

The doubling in total electric generation from 1970 to 1980 is consistent with the long-term trend of a doubling in power consumption every ten years.[10] Beyond 1980, however, a certain degree of saturation in the demand for electricity is likely to occur, so that the demand for it will probably increase at a more modest rate.

THE ELECTRIC UTILITIES' DEMAND FOR COAL

In order to serve the purpose of this analysis, the demand for electricity must be translated into its derived component demands for direct-energy sources, in particular fossil fuels. Table 10 shows the fossil fuel sources of

Figure 1. The Energy/GNP Ratio, 1947-1970

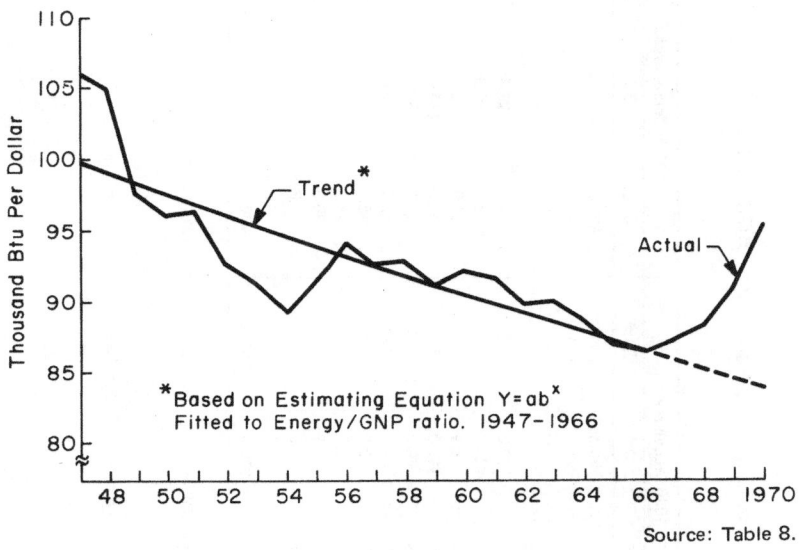

Source: Table 8.

Table 9. National Power Survey Forecast of Total Net Electricity Generation, 1980 and 1985
(gigawatt-hours)

Region	1970[a]	1980	Percent Change	1985	Percent Change
	(1)	(2)		(3)	
New England	60,928	111,000	82.2	152,700	37.6
Middle Atlantic	244,647	452,580	85.0	597,660	32.1
East Central	203,637	376,757	85.1	482,075	27.9
South Atlantic	226,058	448,595	98.4	656,197	46.3
South Central	201,356	483,664	140.2	687,232	42.1
East North Central	209,716	410,760	95.9	556,920	35.6
Central	50,969	136,924	168.6	190,471	39.1
West North Central	48,387	91,530	89.2	127,805	39.6
Mountain	49,256	94,275	91.4	134,975	43.2
Pacific	229,470	476,375	107.6	670,450	40.7
Total[b]	1,524,424	3,082,460		4,256,485	

[a] 1970 preliminary data

[b] Excludes Alaska and Hawaii

Sources: Col. 1: Federal Power Commission, News Release of March 18, 1971. Cols. 2 and 3: Federal Power Commission, *National Power Survey*, Pts. 2 and 3 (Washington, D.C.: Government Printing Office, 1970-1971); printed in Edison Electric Institute, *Fuels for the Electric Utility Industry, 1970-1985*.

Table 10. Forecast of Fossil Fuels Consumed in Electricity Generation, 1970-1985

Region	1970 Consumption			1980 Consumption			1985 Consumption		
	Coal	Oil	Gas	Coal	Oil	Gas[a]	Coal	Oil	Gas[a]
	(Million Tons) (1)	(Million Barrels) (2)	(Million Cfm) (3)	(Million Tons) (4)	(Million Barrels) (5)	(Million Cfm) (6)	(Million Tons) (7)	(Million Barrels) (8)	(Million Cfm) (9)
New England	3.5	75.9	8.2	—	57.8	8.2	—	26.8	8.2
Middle Atlantic	52.7	142.0	176.2	50.0	123.0	176.2	29.0	110.0	176.2
East Central	83.7	2.0	47.9	113.0	0.4	47.9	131.0	0.3	47.9
South Atlantic	57.3	66.9	342.9	48.0	119.0	342.9	51.5	114.0	342.9
South Central	0.8	2.1	1,998.4	10.5	29.0	3,825.0	19.0	39.5	4,437.0
East North Central	84.1	9.3	224.3	109.7	4.1	224.3	112.0	5.8	224.3
Central	15.7	1.0	188.4	47.0	3.5	188.4	48.5	5.6	188.4
West North Central	10.7	1.8	65.9	17.6	2.8	65.9	21.7	4.7	65.9
Mountain	13.1	2.2	171.0	27.0	2.3	226.9	35.5	3.0	252.0
Pacific	0.5	21.8	662.6	19.0	33.5	967.2	22.6	19.0	755.0
Total[b]	322.2	325.0	3,885.8	441.8	375.4	6,072.9	470.8	329.7	6,497.8

[a]For 1980 and 1985 gas volumes are kept constant at the 1970 levels, except in the South Central, Mountain and Pacific regions, see Chapter IV.
[b]Excludes Alaska and Hawaii

Sources: Cols. 1, 2, and 3: Federal Power Commission, News Release of March 18, 1971. Cols. 4 through 9: Federal Power Commission, *National Power Survey*, Pts. 2 and 3.

electric power for 1970-1985, by region. What is interesting is the prediction that beyond 1980 oil consumption by electric utilities will actually decline, while natural gas consumption will remain unchanged except for the South Central, Mountain, and Pacific regions, where either untapped natural gas reserves are in existence or liquified gas could be imported most easily. The inference drawn is that with the exception of these three regions, coal is likely to become an even more important source of fossil-fuel-based electric-power production in the foreseeable future than it is today.

The greatest demand for coal for electricity generation in 1970 originates in the Middle Atlantic, East Central, South Atlantic, and East North Central regions of the United States. By 1985 the forecasts show that the East Central and East North Central regions will dominate the demand pattern for coal shipped to electric utilities. Together, these two regions are predicted to use nearly one-third of the coal shipped to electric utilities. The next largest markets in terms of quantity of coal demanded are the South Atlantic and Central regions.

Kentucky is well situated to serve all these markets, with eastern Kentucky particularly well located to accommodate the East Central and South Atlantic regions by virtue of its proximity to these markets and the availability of water transport to a sizable portion of these two regions.

The alternatives open to electric utilities in choosing a fossil fuel source are basically coal, natural gas, and residual fuel oil. The principal determinants of a utility's long-term fuel choice are the delivered cost per BTU of each fuel, the reliability of delivery to ensure uninterrupted supply, and respective costs of each fuel in meeting clean-air standards. Because adjustment costs of switching from one fuel to another can be large, a temporary change in relative fuel prices is likely to be of less consequence than a major change in such prices that is expected to endure.

Because the cost of shipping coal to distant points is very high, eastern Kentucky coal is a competitive fuel only east of the Mississippi. Likewise, the patterns of interfuel competition are different in this region than in the remainder of the nation. Table 11 shows how fossil fuel demands were met in each of the Bureau of Mines regions in 1972. Striking is the great reliance of New England on oil and gas and the dominant role that coal plays in the East North Central, West North Central, East South Central, and South Atlantic regions.

Of the three fuels, natural gas is the cleanest and the least expensive to burn. Its use in most areas where eastern Kentucky coal is competitive is limited and is primarily on a so-called interruptible basis. This means that the gas supply to a utility may be interrupted when other customers' usage reaches a high fraction of available supplies. Gas is supplied at a lower price to interruptible users than to others, and until recently there were few inter-

Table 11. Steam-Electric Plant Fuel Consumption by Region, 1972

Region	Kilowatt-Hours Generated (Billion)	Percent of BTU Provided by				
		Eastern Kentucky Coal	Other Appalachian Coal	Other Coal	Oil	Gas
New England	52.1	0	0	6	93	1
Middle Atlantic	181.8	0	0	52	44	4
East North Central	290.4	6	3	82	4	5
West North Central	88.8	0	0	62	2	36
South Atlantic	266.6	21	16	25	29	9
East South Central	133.0	13	7	69	2	9
West South Central	198.6	0	0	1	2	97
Mountain	59.4	0	0	58	4	38
Pacific	88.1	0	0	0	30	70
Total U.S.	**1358.8**	**6**	**4**	**44**	**19**	**27**

Sources: National Coal Association, *Steam-Electric Plant Factors*, 1973; and unpublished data, Bureau of Mines.

ruptions. But by early 1974, the incidence of interruptions had increased markedly. This recent higher frequency of interruptions, and its continuation into 1975, is ascribable largely to the pricing policy that governs the natural gas industry.

The price of natural gas sold in interstate commerce is regulated by the United States Federal Power Commission (FPC) and has for some time been kept below market-clearing levels. Instead of promoting balanced development of demand and supply, prices are set at levels which stimulate demand and discourage the development of new supplies.[11] One result has been the increasing occurrence of interruptions of natural gas service to electric utilities and other customers. Although the Congress appears reluctant to deregulate natural gas prices, the FPC is now allowing selective rate increases in the hope of stimulating new supplies but probably at prices per BTU in excess of coal.[12]

The strongest competitor to coal's share in the electric utility market is residual fuel oil. This oil is a thick, tarlike substance which remains when crude oil is distilled into lighter products such as gasoline or diesel fuel. It is generally low in sulfur and less expensive than coal when the costs of handling and storage and that of the boilers using it are considered. Domestic oil refiners have generally found it profitable to produce as little residual oil from each barrel of crude oil as is technologically possible, while foreign refiners have traditionally produced considerable volumes of it.

Until 1966 coal was protected from competition with foreign-produced residual fuel by a set of strictly enforced oil-import quotas. In that year, however, the quota was lifted from oil imports into the region known as Petroleum Administration for Defense District 1, which included the thirteen Atlantic Coast states and Pennsylvania, Vermont, and West Virginia. As a result, oil began to be substituted for coal along the Eastern Seaboard. This trend was strengthened by the enactment of air-quality standards which are particularly stringent in the major East Coast cities. Table 12 lists a comparison of the coal- and oil-market shares in this region in 1966 and 1972. It shows that with the exception of New Hampshire all the areas have significantly increased their use of oil as a fuel source to generate electric power. In most instances, this was accomplished at the expense of using less coal.

The prospect of still tighter sulfur-emission standards at first seemed to assure a further erosion of coal's overall market share. But escalating oil prices in association with the temporary oil embargo of 1973 created some apprehensions in the minds of utility officials. By 1976 some of these apprehensions moderated, but the lack of a comprehensive national energy policy continues to torment utility planners.

A discussion of the alternatives to coal as a fuel for generating electric power would be incomplete without mention of the prospective role of

Table 12. Percentage of BTUs Derived from Coal and Oil
by Electric Utilities along the Eastern Seaboard,
1966 and 1972

State or City	1966		1972	
	Coal	Oil	Coal	Oil
Maine	0	100	0	100
New Hampshire	45	55	68	32
Massachusetts	45	50	1	97
Rhode Island	62	37	0	99
Connecticut	84	16	1	99
New York City	35	46	1	89
New York State (except New York City)	99	0	57	39
New Jersey	61	32	10	83
Philadelphia	76	24	28	71
Pennsylvania (except Philadelphia)	100	0	99	1
Delaware	87	1	45	51
Maryland	99	1	46	51
District of Columbia	98	2	14	86
Virginia	99	0	44	55
North Carolina	99	0	95	4
South Carolina	79	1	84	5
Georgia	100	0	80	8
Florida	20	53	18	58

Sources: National Coal Association, *Steam-Electric Plant Factors*, 1967, 1973.

nuclear energy. Once touted as the solution to escalating needs for electricity, nuclear-power plants today generate less than 5 percent of the electric-power supply.[13] The reasons for the failure of nuclear power to fulfill earlier expectations include controversy over possible adverse environmental effects, rising construction costs, safety, and several technical issues. It takes approximately ten years from the time a decision is made to construct a nuclear plant to the day that plant actually becomes operational.[14] For several years, therefore, and perhaps even decades, the role of nuclear power in generating electricity will be limited to the relatively few plants already operating or under construction. Over the longer term, however, perhaps a generation or more, the situation may be different.

The Battelle Memorial Institute estimates that nuclear power may supply as much as 50 percent of our electric-power needs by the year 2000.[15] Most of this power would come from second-generation nuclear reactors known as "breeder reactors." These reactors are currently under development and are expected by the Atomic Energy Commission "to be able to meet environmental quality and safety standards."[16] The reactors function by converting nonfissionable uranium into nuclear fuel, generating electricity in the pro-

Table 13. Distribution of Bituminous Coal Shipments for Eastern Kentucky and the Remainder of District 8, by Major User, 1970 and 1972
(million tons)

	Eastern Kentucky				Remainder of District 8			
	1970	Percent Share	1972	Percent Share	1970	Percent Share	1972	Percent Share
Electric Utilities	30.0	41.4	43.0	60.7	32.0	36.1	28.3	34.0
Coke Plants	15.4	21.3	13.5	19.0	19.3	21.8	23.4	28.1
Retail Dealers	3.6	4.9	0.8	1.1	3.5	4.0	3.7	4.4
All Other Manufacturing and Industrial Users	13.9	19.2	6.4	9.0	14.4	16.2	16.7	20.0
Overseas Exports	9.6	13.2	7.2	10.2	19.3	21.8	11.2	13.4
Total	72.5	100.0	70.9	100.0	88.5	99.9	83.3	99.9

Note: Percentage may not sum to 100 due to rounding.

Source: Unpublished data, Bureau of Mines.

cess. The impact of this development on coal's electric utility markets could be enormous. However, this development has only long-term prospects. At this point it is beset with all the uncertainties that attach to very long forecasts. Serious reservations have been raised in the past two years concerning the wisdom of building and using large-scale nuclear-power plants, and at the present time the issues remain yet to be resolved.

Since eastern Kentucky coal is low in sulfur and high in BTU content and since it traditionally has not been hampered by supply interruptions as much as coal mined elsewhere, demand for it has been strong in recent years. Electric utilities have actively bid for it for several years. Table 13 shows the impact of this situation on shipments of eastern Kentucky coal to electric utilities. Eastern Kentucky producers shipped nearly two-thirds of their 1972 production to electric utilities, a 50 percent increase over two years. [17] When compared with the rest of District 8 shipments, this is 78.5 percent greater than the one-third share electric utilities hold in the market. The capricious nature of oil markets, along with an environmentally conscious world, have made eastern Kentucky coal a very attractive resource commodity.

In addition to competing with other coals on the basis of chemical property, eastern Kentucky coal also competes with coals on the basis of location. This is, of course, due to the impact of distance on transportation costs, hence on delivered price. In 1972, for example, electric utilities in New York State consumed nearly six million tons of coal. More than one-half of this

Table 14. Shipments of Coal from Eastern Kentucky to
Electric Utility Plants, 1972
(thousands of net tons)

	Shipments	Percent of Total Eastern Kentucky Shipments to Utilities	Percent of Total Utility Shipments Received
Middle Atlantic			
New Jersey	50	0.1	4.0
New York	125	0.3	2.2
Total	175	0.4	
South Atlantic			
Delaware and Maryland	460	1.1	8.5
District of Columbia	75	0.2	51.4
Virginia	4,340	10.1	88.7
West Virginia	2,230	5.2	9.8
North Carolina	9,950	23.1	50.5
South Carolina	2,325	5.4	42.4
Georgia and Florida	6,110	14.2	35.2
Total	25,490	59.2	
East North Central			
Indiana	895	2.1	3.4
Illinois	775	1.8	2.4
Michigan	2,320	5.4	10.8
Ohio	3,318	7.7	7.9
Wisconsin	365	0.8	3.4
Total	7,673	17.8	
East South Central			
Alabama	152	0.4	0.8
Kentucky	3,375	7.8	14.4
Tennessee	5,375	12.5	28.5
Total	8,902	20.7	
West North Central			
Iowa	750	1.7	13.8
Minnesota	40	0.1	0.6
Total	790	1.8	
GRAND TOTAL	43,030	99.9	

Sources: Bureau of Mines, "Bituminous Coal and Lignite Distribution, Calendar Year 1972," *Mineral Industry Surveys,* 1973; and unpublished data, Bureau of Mines.

coal originated in Pennsylvania, the closest coal-producing state, another third came from Districts 3 and 6 (northern West Virginia) and less than 5 percent originated in District 8. It is not that coal from the other areas was of better quality than District 8 coal, but rather that its delivered price was lower because of lower transportation costs.

The most important Bureau of Mines region for eastern Kentucky coal shipments is the South Atlantic region. In 1972 the states in this region absorbed nearly two-thirds of eastern Kentucky's output with North Carolina alone buying nearly ten million tons or almost one-fourth of eastern Kentucky's output. As Table 14 shows, the second most important region is the East South Central region, which includes Tennessee. The third most important set of state markets for eastern Kentucky coal, Georgia and Florida, together bought 14.2 percent of eastern Kentucky's output.

Substantial shipments are also made to parts of the East North Central region, despite strong competition from coal mined in the Eastern Interior Coal Basin. Ohio, which consumes as much eastern Kentucky coal as Kentucky itself, receives more than one-half of this coal via river barge, by far the least expensive method of transporting coal.

Map 4 shows that by far the largest share of eastern Kentucky coal is shipped northeast or southeast, with but a fraction going north and nothing at all west. The major market for this coal lies south of Kentucky, accounting for 55.6 percent of total output. Given the rapid economic growth of this region of the United States, it can be expected that, other things equal, the electric utility demand for eastern Kentucky coal will also expand rapidly.[18]

A state ranking in terms of the amount of eastern Kentucky coal which electric utilities located in each state consumed in 1972 is shown in Table 15 (see column 1). Column four of that table shows that one-half of the shipments went to four states, North Carolina, Georgia, Florida, and Tennessee. Even though three of these states are not contiguous to Kentucky, eastern Kentucky coal does represent a major fuel source for them.

Overall, buyers of eastern Kentucky coal are fairly well dispersed with regard to shipments to electric utilities. Ninety-one percent of the output destined for public utility consumption is distributed over ten states with none absorbing more than 23 percent of the output for electric utilities. As shown in column four, the remaining 9 percent are distributed over eleven states with none purchasing more than 2 percent.

The case of electric utility's dependence on eastern Kentucky for coal supplies is somewhat different. Striking is the fact that 88 percent of the coal shipments received by electric utilities in Virginia originate in eastern Kentucky, a very high dependency. North Carolina, South Carolina, Georgia, and Tennessee are also strongly dependent upon eastern Kentucky coal for their coal-fueled electric power (see column 3, Table 15). The same can be said for

Map 4. Regional Distribution of Eastern Kentucky Coal Shipments to Electric Utilities
(thousands of net tons and percentage)

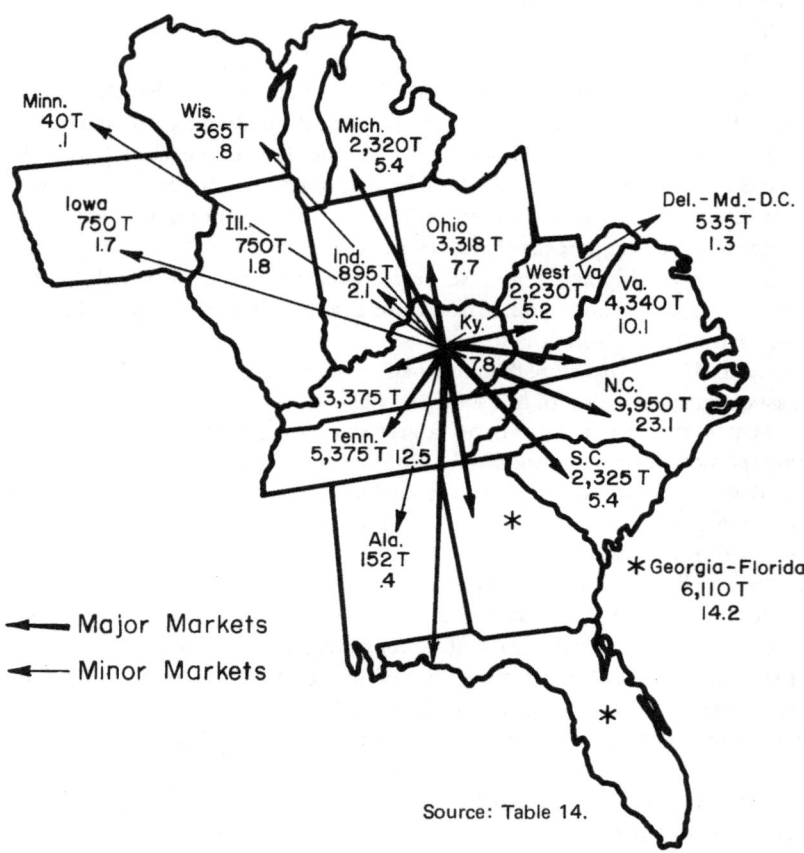

Source: Table 14.

the District of Columbia, but total consumption is quite small, a mere 75,000 tons.

It is quite obvious that a forecast of the electric-utility demand for eastern Kentucky coal for more than a year or two ahead involves a great many imponderables. Among these are the relative prices of coal and oil, the severity of required air-pollution standards and their enforcement, the availability and cost of equipment for the removal of pollutants from utility smokestack gases, and the rate of growth in demand for electric power in those states that consume large amounts of eastern Kentucky coal.

Table 15. Rank-Size Distribution of Electric Utility
Market for Eastern Kentucky Coal, 1972
(thousands of net tons and percentage)

State	Rank	Tons	Percent of Total	Cumulative Percent
North Carolina	1	9,950	23.1	23.1
Georgia and Florida	2	6,110	14.2	37.3
Tennessee	3	5,375	12.5	49.8
Virginia	4	4,340	10.0	59.8
Kentucky	5	3,375	7.8	67.6
Ohio	6	3,318	7.7	75.3
South Carolina	7	2,325	5.4	80.7
Michigan	8	2,320	5.4	86.1
West Virginia	9	2,230	5.2	91.3
Indiana	10	895	2.1	93.4
Illinois	11	775	1.8	95.2
Iowa	12	750	1.7	96.9
Delaware and Maryland	13	460	1.1	98.0
Wisconsin	14	365	0.8	98.8
Alabama	15	152	0.4	99.2
New York	16	125	0.3	99.5
District of Columbia	17	75	0.2	99.7
New Jersey	18	50	0.1	99.8
Minnesota	19	40	0.1	99.9
		43,030		

Source: Calculated from Table 14.

In principle, several sets of forecasts of national coal demand could be disaggregated, and on the basis of selected demand parameters, an appropriate portion could be designated to eastern Kentucky. But a perhaps more useful alternative is to rely on the recent study of the Appalachian coal industry by Charles River Associates (CRA).[19] The study includes a set of 1980 forecasts of utility coal consumption by states east of the Mississippi, based on a number of scenarios of future relative prices and air-pollution standards. When one of these scenarios is combined with the 1972 data on eastern Kentucky's share in each state's electric utilities coal market, a 1980 forecast of electric-utility demand for eastern Kentucky coal can be estimated.[20]

Table 16 shows CRA's forecast of 1980 utility coal consumption in each state that uses eastern Kentucky coal. An underlying assumption embedded in this forecast is that no coal markets are lost to residual oil during 1970-1980. The eastern Kentucky total of coal shipments to these states is based on market shares extant in 1972.

It might be argued that in light of the quantum leap of oil prices in 1974 these forecasts, because they were prepared before 1974, are no longer useful. However, it is important to remember two things. One, the steep rise in coal prices that accompanied the rise in oil prices in 1974 left relative

Table 16. 1980 CRA Forecasts of Coal Consumption by Electric Utilities and Resulting Eastern Kentucky Demand Forecasts
(millions of tons)

	(1)	(2a)	(2b)	(3a)	(3b)
Middle Atlantic					
New Jersey	3.80	0.00	2.13	0.00	0.00
New York	14.78	9.18	14.73	5.65	10.18
South Atlantic					
Delaware and Maryland	11.83	1.56	5.76	1.19	3.24
District of Columbia	0.48	0.48	0.48	0.00	0.07
Virginia	8.48	3.22	6.17	1.08	5.71
West Virginia	18.53	18.23	18.53	12.12	18.23
North Carolina	38.59	36.44	38.59	15.97	31.33
South Carolina	2.58	2.31	2.54	0.00	1.37
Georgia and Florida	17.01	11.24	16.51	0.00	11.45
East North Central					
Indiana	52.07	52.07	52.07	43.87	52.07
Illinois	68.29	68.29	68.29	59.74	68.29
Michigan	41.43	41.43	41.43	35.64	41.43
Ohio	67.08	58.73	67.08	34.11	66.81
Wisconsin	14.72	14.63	14.78	14.48	14.63
East South Central					
Alabama and Mississippi	32.70	32.70	32.70	18.56	32.70
Kentucky	25.95	25.95	25.95	23.56	25.95
Tennessee	10.49	10.49	10.49	9.31	10.49
Eastern Kentucky Shipments	58.29	48.56	55.47	26.59	48.45

(1) No coal markets lost to residual oil, 1970-1980.
(2) Coal price rises by three cents per million BTU (75 cents per ton) over 1969 relative prices and (a) Residual fuel prices remain at their 1969 P.O.E. prices and inland transport costs are *low* (1 mill per barrel mile). (b) Residual fuel prices rise 25 cents per barrel over 1969 prices and transport costs are *high* (2 mills per barrel mile).
(3) Same as (a) and (b) with enforcement of 1975 air quality standard of 0.7 percent sulfur-in-fuel standard for coal.

Source: Charles River Associates, *Economic Impact of Public Policy.*

prices basically unchanged. Second, through the mechanism of the "fuel cost adjustments," public utilities have been able to pass on to their customers higher oil prices in the form of higher electricity rates. And the consumer demand for electricity remains relatively stable. In addition, oil is once again available in great abundance on the world market.

Consequently, electric utilities have generally continued to use oil as their primary fuel, and coal's share in this market has not expanded as rapidly as one might have thought in looking only at the price dislocations of the past two years. Public utilities seem to view oil as more flexible than coal and less likely to violate clean-air standards.

Some utilities, however, in particular the Tennessee Valley Authority, have begun to encourage the taking of bids for long-term contracts from low-sulfur coal producers in eastern Kentucky. In analyzing their long-range plans, some company officials remain uncertain about the best way to meet air-pollution standards—by using more expensive low-sulfur coal or by installing sulfur-purging equipment at their plants. As of this time, the use of low-sulfur coal is being encouraged, but whether or not this decision will endure is difficult to predict.

The CRA estimates are best viewed as highly conservative, but useful as a bench mark for updated predictions. It is clear that the size of future coal shipments from eastern Kentucky to electric utilities will depend largely on the price of this coal in relation to competing low-sulfur coal, on the price of oil, and on the state of enforcement of air-pollution standards. On the whole, prospects for eastern Kentucky producers of low-sulfur coal are favorable.

THE DEMAND FOR COKING COAL

The production of coal for domestic coking purposes is the second largest mining activity in eastern Kentucky, accounting for 19 percent of total shipments in 1972. Coking processes require a high-quality coal with low sulfur and ash content. Much of eastern Kentucky coal meets or exceeds these requirements and, therefore, is sought for coking purposes. Some of this high-quality coal is also exported to coking plants in Canada, Europe, and Japan.

There are three principal factors that influence the aggregate volume of coking coal demanded per ton of steel produced.[21] The first of these is the volume of coal required per ton of coke. While this ratio can vary widely, depending upon the particular coal used, it is typical for several different coals to be blended for use in the production of coke. The apparent effect of this practice in the postwar period has been to achieve a uniform blend of coking coal, for the ratio has been stable during the period between 1.42 and 1.45 tons of coal per ton of coke produced.

A second factor is the volume of coke used per ton of pig iron. This ratio has been declining, although somewhat erratically, over the postwar period. Technological changes have reduced the coke-to-pig-iron ratio from 92 percent in 1945 to 61 percent in 1972. Furthermore, development is under way on a process which would entirely eliminate the need for coke in the production of pig iron. This new process of pelletizing iron ore for use in an electric furnace, however, is not yet proved commercially. It is not likely to have much of an impact for some time. Whether the decline in the coke-to-pig-iron ratio will continue without the use of the new process is uncertain.

The final factor is the amount of pig iron consumed per ton of steel produced. Although steelmaking technology has changed dramatically over the postwar period, the pig-iron-to-steel ratio has remained stable at between 64 and 70 percent. The key to this stability is the availability of scrap steel. Most scrap steel is not derived from old steel products but is a new by-product of the steel production process itself. Its availability is therefore roughly pro-

Table 17. Relative Share of Coking Coal Production,
1970 and 1972
(percentage)

Coking Coal Production as a Fraction of Total Bituminous Coal produced in	1970	1972
United States	16.0	14.9
Eastern Kentucky	21.3	19.0
Remainder of District 8	21.5	21.8

Sources: Unpublished data, Bureau of Mines; and *Minerals Yearbook,* 1973.

portional to the amount of steel produced. Thus, although scrap steel is partially substitutable for pig iron in steel production, its proportional availability implies relative constancy in the pig-iron-to-steel ratio. This ratio may change occasionally, in response to temporary aberrations in the relative prices of scrap steel and pig iron. But, given the strong worldwide market for scrap steel, there is usually little old scrap available to replace pig iron in steel production.

A forecast of the aggregate demand for coking coal depends upon these three factors and upon the projected level of steel production. In the past, it would have been reasonable to expect steel production to rise more or less proportionately with real GNP. For the next several years, however, implicit import restrictions and the tight worldwide steel market may mean that U.S. steel producers could garner a greater share of the domestic market than they have in the past.[22]

The low-sulfur content which has made eastern Kentucky coal so well suited to coking purposes is also important to electric utility customers of the region's mines. This fact, of course, reflects the steady tightening of air-quality standards over the past several years. The remaining question is whether eastern Kentucky will increase its sales of coking coal in proportion to the national total or will become more specialized as a supplier of low-sulfur fuel for electric utilities. The behavior of coal shipments during the 1970-1972 period would seem to lend weight to the second possibility. As may be seen in Table 17, the relative importance of eastern Kentucky sales changed in the 1970-1972 period in about the same way as national sales. Production in the remainder of District 8 ran counter to the national and eastern Kentucky patterns. This means that for District 8 (exclusive of eastern Kentucky) no change in user share of coal output occurred during

Table 18. Forecasts of 1980 Domestic Coking Coal Shipments from the U.S. and Eastern Kentucky
(thousands of tons)

	Low Growth of Steel Production	High Growth of Steel Production
Total United States	103,345	115,953
Eastern Kentucky — with 12 Percent of National Market	12,401	15,501
Eastern Kentucky — with 15 Percent of National Market	13,914	17,393

Source: Charles River Associates, *Economic Impact of Public Policy.*

this period. But in eastern Kentucky, coal shipped to electric utilities expanded its share in coal output while coking coal's share declined.

Many unknown factors are involved in forecasting the demand for coking coal; therefore a range of estimates for 1980 based upon several assumptions must be interpreted with care. The estimates are shown in Table 18, where two forecasts of 1980 national coking coal demand are given. These are based, respectively, upon a "low" (3.5 percent) rate of growth and a "high" (5 percent) rate of growth of steel production between 1972 and 1980. They are also based upon a coal-coke ratio of 1.43, a coke-pig-iron ratio of 0.54, and a pig-iron-steel ratio of 0.67. Of these, only the coke-pig-iron ratio is set differently from its recent average. This ratio has been falling steadily since World War II and will probably continue to decline into the 1980s.

Two eastern Kentucky alternatives are offered at each rate of national coking-coal production. These depend upon eastern Kentucky's supplying either 14 percent or 10 percent of aggregate national production. It is probable that the resultant four projections bracket the approximate future levels of demand for coking coal from eastern Kentucky.

Other industrial demands for coal, although quite important to District 8 and eastern Kentucky, are steadily declining sources of coal sales for the region. Most of the coal sold in this category is used to produce industrial "process steam," which is needed for industrial processes themselves rather than for the generation of electric power. Nearly 15 percent of 1972 District 8 coal shipments and about 9 percent of eastern Kentucky shipments belong in this category, but the volume of such shipments from District 8 has been declining in recent years at an annual rate of about 6 percent. There was a particularly sharp drop in eastern Kentucky shipments between 1970 and 1972, but it is supposed that eastern Kentucky will, in the future, retain its 1972 share of District 8 shipments.

The demand for coal for industrial use varies, of course, with the general level of economic activity. It can be assumed that industrial coal consumption will continue to decline at the long-term average annual rate of 6 percent and that any single year's consumption will be the product of this trend and the current state of the economy. District 8 coal shipments to industrial users would then decline by 1980 by an additional 40 percent. If so, and assuming that eastern Kentucky retains its 28 percent share of the District 8 market in this category, eastern Kentucky would ship 3.8 million tons to industry in that year.

THE RETAIL DEMAND FOR COAL

Retail sales of coal, primarily for residential heating, were once an important component of coal demand. Since World War II, however, most home owners have converted their furnaces to other fuels, and retail sales of coal have fallen steadily.

This category of coal demand represented only about 3 percent of District 8 production and 4 percent of eastern Kentucky production in 1972. A forecast of eastern Kentucky shipments is based on the assumption that the eastern Kentucky share of District 8 will remain at its 1972 level. Total District 8 shipments fell at an annual rate of 14 percent between 1968 and 1972. If they continue to decline, by 1980 District 8 will sell 1.3 million tons for retail delivery. This compares to 4.4 million tons sold in 1972. If eastern Kentucky producers retain their 1972 share in District 8 retail deliveries—17.5 percent—they will sell approximately 230 thousand tons in 1980.

THE EXPORT DEMAND FOR COAL

Only the nation's highest quality and most valuable coal is competitive on world markets. About half of U.S. coal exported to Canada is destined for electric utility plants. But the other half, as well as all overseas shipments of coal, is destined for coking purposes. Most exported coal is from Appalachia, and, excluding shipments to Canadian electric utilities, about half of that comes from District 8 (see Table 19). Eastern Kentucky exports less of its output than the remainder of District 8, probably because of West Virginia's relative proximity to the main shipping port of Norfolk and favorable transport rates.

Coal exports to Canada are a smaller percentage of the world market for U.S. coal than are its overseas exports. Transportation costs to Canada, the level of economic activity there, and the relative costs of western Canadian and U.S. coal determine this component of export volume. While coal from

Table 19. Coal Exports from the U.S. and District 8, 1972
(thousands of tons)

	United States	District 8
Exports to Canada Electric Utilities	8,821	30
Other Exports to Canada	8,919	5,633
Overseas Exports	36,607	18,422

Source: Bureau of Mines, "Bituminous Coal and Lignite Distribution, Calendar Year 1972," *Mineral Industry Surveys,* 1973.

Western Canada is suitable for coking, there is a strong market for the coal in Japan. Thus, it is problematic whether coal from Western Canada can displace any District 8 coking coal sold in Eastern Canada. In any event, Canadian shipments usually represent only about one-fourth of District 8 exports, and eastern Kentucky usually accounts for only about 30 percent of District 8 exports. Even a large decline in U.S. coking-coal exports to Canada therefore would not affect eastern Kentucky significantly.

Overseas exports from the United States flow primarily to Europe and Japan. Historically, exports to Europe have been determined not only by the level of economic activity there but also by Common Market policy concerning coal imports. An example of this was the establishment of subsidies to the European coking-coal industry by the European Coal and Steel Community in the mid-1960s. Lately, Eastern Europe has increased its coal shipments to the Common Market countries, and U.S. coal exports to Europe have fallen.

U.S. coal exports to Japan are not subject to import limitations and are determined by competitive world conditions and Japanese steel output. Thus, when Japanese steel production doubled between 1966 and 1970, U.S. coal shipments to Japan rose from 7.8 million to 27.6 million tons.[23] Our transport costs are high, however, relative to those for shipments from Western Canada and Australia. These nations are supplying, through long-term contracts, an increasing proportion of Japanese coal demand. They are expected to supply from 50 to 70 percent of those needs by 1980.[24]

In the CRA study of the Appalachian coal industry, a range of forecasts of U.S. coal exports for 1980 is presented. These forecasts, summarized in Table 20, are the basis for the forecasts of eastern Kentucky coal exports. The procedure used is as follows: It is assumed that District 8 retains its 1972 share in U.S. exports to Canada and overseas countries and that eastern Kentucky producers retain their 1972 share in District 8 exports. The resulting forecasts are shown on the bottom line of Table 20.

Table 20. 1980 CRA Forecasts of Coal Exports from the U.S. and Resulting Eastern Kentucky Forecasts
(million tons)

U.S. Exports to	Low	Middle	High
Canada	19	22	24
Common Market	20	27	35
Japan	19	36	51
Others	10	15	19
Total	68	100	129
Eastern Kentucky Exports	9	14	18

Source: Charles River Associates, *Economic Impact of Public Policy.*

So far, this chapter has explored the determinants of the demand for eastern Kentucky coal by major category of user. It was established that electric utility demand for the region's coal is not only the dominant but also the fastest-growing component of total demand. Demand for coking coal and from export markets is likely to grow, too, but more moderately. The remaining categories, retail and other manufacturing and industrial uses, are likely to continue their long-term decline.

Table 21 presents data on actual 1972 shipments and the mid-range 1980 forecast of shipments from eastern Kentucky for each major user category. The total of these demand components is 91.9 million tons, as compared to an actual figure of 70.9 million tons shipped in 1972. It may be reasonable to add another 10 million tons to the forecast of shipments to electric utilities, bringing the total to 102 million tons. According to Table 21, this addition could come about if no eastern Kentucky coal markets are lost to oil by 1980. That may well turn out to be the case.

THE SUPPLY OF EASTERN KENTUCKY COAL

The supply of coal offered for sale by mining firms depends on a series of cost, market, and institutional factors, all of which are in a constant state of flux. These factors and their influence on the flow of eastern Kentucky coal to markets are discussed here.

As Table 5 shows, the Appalachian Coal Basin supplied well over half of the nation's 1974 output of coal. The four largest coal-producing states alone contributed 51 percent of that output. If these four states are examined individually, it becomes clear that the most vigorous growth in coal production over the last six years has occurred in eastern Kentucky. From 1969 to 1974,

Table 21. Actual 1972 Shipments and Forecast
of 1980 Shipments of Coal from Eastern Kentucky
(thousands of tons)

	1972 Actual	1980 Forecast
Electric Utilities	43,030	58,290[a]
Coke Plants	13,480	15,650[b]
Retail Dealers	770	230
All Other Manufacturing and Industrial Uses	6,393	3,800
Exports	7,230	14,000
Total	70,903	91,970

[a]Based on alternative (1) from Table 16. [b]Based on a moderate rate of growth of steel production and a 15 percent share of the coking coal market for eastern Kentucky.
Sources: Table 13 and Charles River Associates, *Economic Impact of Public Policy*.

coal production in the eastern part of the state has increased by one-fourth, while Pennsylvania's output remained constant and Ohio's and West Virginia's declined significantly (see Table 22). In fact, the latter's annual output declined in percentage terms by an amount equal to that by which eastern Kentucky's output increased. It would seem, therefore, that the Appalachian Coal Basin's ability to supply a vigorous demand for coal rests entirely on eastern Kentucky's ability to continue to mine at an accelerated pace. During the years 1970 to 1972, output in eastern Kentucky declined slightly, only to resurge in 1973 and 1974 to record levels of production.

The source of this rise in output is found in the surface-mining sector of the industry. As Table 23 shows, underground mining declined sharply between 1969 and 1974 in the four major regions of the Appalachian Coal Basin, largely as a result of the passage of the 1969 Coal Mine Health and Safety Act. Because of altered relative costs of mining, surface mining became during these years a much preferred alternative to underground mining. Production statistics show that while underground mining declined by 26 percent, surface mining increased by 41 percent. But the increase in surface mining in the basin was not sufficient in absolute tonnage to offset entirely the large decrease in underground mining. As a consequence, total output for the four major producing states in the basin declined by 8 percent.

The rapid growth in surface mining and the decline in underground mining are best illustrated by the respective shares of each in total coal production. In 1969 surface mining represented only 28 percent of total coal

Table 22. Coal Output for Four Appalachian Coal Basin States, 1969-1974
(million short tons)

	1969	1970	1971	1972	1973	1974	Percentage Change
West Virginia	141.0	144.1	118.3	123.7	115.4	106.0	−24.83
Eastern Kentucky	61.6	72.5	71.6	68.9	74.0	77.1	+25.15
Pennsylvania	78.6	80.5	72.8	75.9	76.4	78.9	+ 0.38
Ohio	51.2	55.4	51.4	51.0	45.8	44.6	−12.90

Sources: Bureau of Mines, "Coal—Bituminous and Lignite," *Minerals Yearbook*, 1969-1974.

Table 23. Production of Surface and Underground Mined Coal in West Virginia, Eastern Kentucky, Pennsylvania, and Ohio, 1969-1974
(million short tons)

	1969	Percent Share	1974	Percent Share	Percent Change
Underground	240.8	72	177.5	58	− 26
Surface Mining	91.7	28	129.0	42	+ 41
Total	332.5	100	306.5	100	− 8

Source: Calculated from Table 24.

mined in the four regions of the basin (see Table 23). By 1974 surface mining had increased its share to 42 percent, while underground mining's share had declined to 58 percent.

The decline in output from the Appalachian Coal Basin would have been much more severe if production in eastern Kentucky surface mining had not risen so dramatically. As Table 24 shows, surface output in eastern Kentucky more than doubled in the six-year period, for an average annual increase of 24 percent. West Virginia's and Pennsylvania's output also rose substantially, but far less rapidly than that of eastern Kentucky.

The rapid increase in surface mining in eastern Kentucky is largely explained by the ease of entry that characterizes this sector of the industry. Reclamation requirements, the posting of bonds, frequency of inspection, and other obstacles to freedom of entry, that is, to the establishment of new surface mines or the expansion of existing ones, are less severe in eastern Kentucky than in neighboring states. Typically, a vigorous demand translates into rising prices, which in turn give an incentive to small operators to enter surface mining.

Table 24. Production of Coal in Four Appalachian Coal Basin States, Underground and Surface, 1969-1974
(million short tons)

	1969	1970	1971	1972	1973	1974[a]	Percentage Change
West Virginia							
Underground	121.6	116.4	92.4	101.7	95.5	82.0	−32.57
Surface	19.4	27.7	25.8	22.1	19.9	24.0	+23.71
Pennsylvania							
Underground	56.0	55.4	44.3	49.1	46.2	42.0	−25.00
Surface	22.6	25.1	28.5	26.8	30.2	36.9	+63.27
Eastern Kentucky							
Underground	44.5	43.2	37.4	37.9	40.6	39.5	−11.24
Surface	17.1	29.3	34.2	30.9	33.4	37.6	+119.88
Ohio							
Underground	18.6	18.1	12.9	16.3	16.2	14.0	−24.74
Surface	32.6	37.2	38.6	34.7	29.6	30.5	−6.45
Total	332.5	352.4	314.1	319.5	311.6	306.5	

[a]Preliminary estimates

Sources: Bureau of Mines, "Coal—Bituminous and Lignite," *Minerals Yearbook,* 1969-1974.

The increasing importance of surface mining relative to that of underground mining is illustrated in Table 25. During the past six years eastern Kentucky surface mines have steadily increased their share in total output. At the same time, the underground mines' share in total output has declined so that in 1974 total output was shared almost equally between underground and surface mining.

In Pennsylvania, the relative decline in underground mining in favor of surface mining has been similar to that of eastern Kentucky. In Ohio and West Virginia, on the other hand, surface mining has increased its share of output much more slowly. In the former, surface mining has always dominated coal production for topographical reasons. The terrain is flatter in Ohio than in the neighboring states, and surface mining has been traditionally an economically preferred method of mining. Surface mining in West Virginia, with stringent reclamation requirements, has increased somewhat more slowly than neighboring eastern Kentucky. The trend toward relatively more surface mining, however, is unmistakable, as it is in eastern Kentucky and Pennsylvania. But with less than one-fourth of West Virginia's annual output originating from surface operations, underground mining continues to dominate output. Because of the much longer lead times required to open

Table 25. Percentage Distribution of Coal Mining by Type of Mine for Four Appalachian Coal Basin States, 1969-1974

	Year	Underground Percent of State	Surface Percent of State
Eastern Kentucky	1969	72	28
	1970	60	40
	1971	52	48
	1972	55	45
	1973	55	45
	1974	51	49
Ohio	1969	36	64
	1970	33	67
	1971	25	75
	1972	32	68
	1973	35	65
	1974	33	68
Pennsylvania	1969	71	29
	1970	69	31
	1971	61	39
	1972	65	35
	1973	60	40
	1974	53	47
West Virginia	1969	86	14
	1970	83	17
	1971	78	22
	1972	82	18
	1973	83	17
	1974	77	23

Sources: Bureau of Mines, "Coal—Bituminous and Lignite," *Mineral Industry Surveys,* 1969-1974.

new or to expand existing underground mines, growth in output in response to outside stimuli is likely to be much more restrained in West Virginia than in Ohio or eastern Kentucky.

An additional factor that influenced the increase in eastern Kentucky coal production is the low-sulfur content of its coal. Table 26 shows the characteristic ranges of coal deposits in the four subregions. Noteworthy are the narrow sulfur range of eastern Kentucky coal and its high minimum BTU content. This is very attractive coal indeed in the light of air-pollution restrictions and the desire to minimize per BTU costs.

THE COAL RESERVE BASE

The recent concern with preserving a clean environment has focused attention on the burning of low-sulfur coal. Regardless of whether Environ-

Table 26. Coal Characteristics in Four Appalachian Coal Basin States

	Rank	Percent Content Ash	Sulfur	Heat Content BTU/lb
Eastern Kentucky	High-volatile bituminous Some good coking coal	3.5-11.0	0.58-3.3	12,300-14,200
West Virginia	High-volatile bituminous Much good coking coal	2.0-28.0	0.40-14.0	10,200-15,600
Pennsylvania	High-volatile bituminous and anthracite and scrui-anthracite, much good coking coal	4.1-15.1	0.50-4.2	12,580-14,490
Ohio	High-volatile bituminous	6.6-11.5	1.5-5.0	11,006-12,919

Source: Committee on Science and Astronautics, *Energy Facts.*

mental Protection Agency (EPA) standards are implemented in 1977 or delayed, low-sulfur coal appears to be in ever-increasing demand. If the EPA standards are implemented as now scheduled, the demand for low-sulfur coal will accelerate even more rapidly than it already has. Thus, the question of adequacy of coal reserves takes on greater importance.

Surveys conducted by the Bureau of Mines and the Geological Service reveal that approximately 83 percent of the bituminous coal and nearly all the anthracite deposits are located east of the Mississippi River, while subbituminous and lignite coal deposits are found west of the river.

Table 27 shows the size of the bituminous reserves in the four major producing regions of the basin. Eastern Kentucky underground reserves appear a distant fourth to the rather abundant deposits in West Virginia, Pennsylvania, and Ohio, while surface deposits compare favorably with those of the other three regions.

If it is assumed that 50 percent of underground and 90 percent of surface coal deposits are recoverable, then, given production rates, the life of the demonstrated reserve base can be estimated.[25] Table 28 shows the estimated life for underground and surface mining on the assumption that 1974 annual production rates will continue unchanged. The table reveals that, with the exception of minable surface deposits in Pennsylvania, eastern Kentucky reserves have the shortest life span of the four regions. Surface reserves can last only eighty-four years and underground deposits, 120 years. These reserve lives are considerably shorter than, for instance, the underground Ohio deposits (210 years) and the surface reserves in West Virginia (195 years).

The estimates of reserve lives are sensitive, first, to rates of production, second, to coal-recovery rates, and, third, to the accuracy of the estimate of the reserve base. New discoveries could expand the base, while prohibitions of surface mining on steep slopes could reduce it. Little room remains to

Table 27. Demonstrated Bituminous Coal Reserve Base for Four Appalachian Coal Basin States, January 1, 1974[a]
(million short tons)

	Potential Mining Method		
	Underground	Surface	Total
West Virginia	34,378	5,212	39,590
Pennsylvania	22,789[b]	1,181	23,970
Ohio	17,423	3,654	21,077
Eastern Kentucky	9,467	3,450	12,917

[a] Includes measured and indicated categories as defined by USBM and USGS and represents 100 percent of the coal in place.
[b] Exclusive of anthracite

Source: Bureau of Mines, "Demonstrated Coal Reserve Base of the United States on January 1, 1974," *Mineral Industry Surveys*, 1974.

improve the 90 percent recovery rate in surface mining, but technological advances can and have in recent years improved the recovery rate in underground mining.

It would be useful to be able to calculate how much of coal output originates from the variously angled slopes currently being surface mined. However, no data linking tonnage produced to mountain slope exist. In order to estimate the approximate impact of possible prohibitions against surface mining on steep slopes, the Council on Environmental Quality conducted a sample survey in the Appalachian Coal Basin. Table 29 shows the results obtained by the council for the major producers in the basin. The strong reliance on steep-slope surface mining in eastern Kentucky and West Virginia is striking. More than 85 percent of the former's surface output and 72 percent of the latter's surface originate from slopes in excess of 20 degrees. Pennsylvania and Ohio outputs, on the other hand, emanate from more moderately sloped land. It is clear that a ban against surface mining on slopes in excess of 20 degrees would most severely affect eastern Kentucky and West Virginia, while Ohio and Pennsylvania would be affected practically not at all.

The location of coal reserves also does not favor eastern Kentucky. As Table 29 shows, 75 percent of its strippable reserves are located on slopes in excess of 20 degrees, as compared to 43 percent for West Virginia. Accordingly, a ban on steep-slope surface mining, those in excess of 20 degrees, would render the major portion of eastern Kentucky's surface reserves unsuitable for further mining. The impact upon the economy of that region of the state would undoubtedly be quite severe.

Table 28. Estimated Life of Demonstrated Underground and Surface Coal Reserves in Four Appalachian Coal Basin States at 1974 Production Rates

	Underground Years[a]	Surface Years[b]
Eastern Kentucky	120	84
West Virginia	210	195
Ohio	620	108
Pennsylvania	271	29

[a] Assumed underground recovery rate 50 percent

[b] Assumed surface recovery rate 90 percent

Sources: Calculated from Table 27 and 1974 production statistics, Bureau of Mines, *Weekly Mineral Industry Surveys,* 1975.

Despite the recent surge in the amount of coal mined and in its price, eastern Kentucky remains an economically depressed area highly dependent on coal mining for its existence. Even though efforts have been made in recent years to diversify the economic base of the region, unemployment and poverty levels remain high and per-capita income lags considerably behind the national average.[26]

Most of the coal mined in eastern Kentucky contains low amounts of sulfur (see Table 26). Table 30 shows the percentage distribution of coal reserves by low-, medium-, and high-sulfur content. Even though eastern Kentucky's reserves are smaller than those of its neighbors, a far greater percentage of these reserves has a low-sulfur content. This type of coal is, of course, in high demand today. Only West Virginia coal reserves have a comparable low-sulfur content. Ohio coal is mostly of the high-sulfur type, and remarkably few of its deposits fall into the low to middle sulfur-content class. Pennsylvania as well has a very small amount of low-sulfur coal. The high rate and recent growth of coal production in eastern Kentucky can, at least in part, be explained by the strong demand for low-sulfur coal.

When strippable coal reserves are considered separately, a similar picture emerges with regard to low-sulfur coal. The distribution of strippable coal reserves classified in accordance with low-, middle-, and high-sulfur content shows that relatively plentiful deposits of low-sulfur coal are found only in eastern Kentucky and West Virginia (see Table 31). Pennsylvania's and Ohio's reserves are primarily of the high-sulfur type. The reserves shown in Table 31 were calculated by applying the latest available estimates of sulfur distribution to the strippable reserve base estimates of the four major sub-

Table 29. Surface Mine Production in Four Appalachian Coal Basin States as a Function of Slope Angle, 1971
(million tons)

Region	Total	0° to 9.9°	10° to 14.9°	15° to 19.9°	20° to 24.9°	25° +
				Production		
Eastern Kentucky	33.10	0.00	0.60	4.20	7.65	20.65
Ohio	38.11	4.08	8.00	15.08	6.39	4.56
Pennsylvania	25.76	10.73	9.89	3.63	1.04	0.48
West Virginia	31.92	1.83	4.43	2.71	8.43	14.46
				Strippable Reserves		
Eastern Kentucky	766.6	44.8	38.8	106.4	219.4	357.2
Ohio	1,333.7	961.0	256.4	102.9	13.4	0.0
Pennsylvania	1,293.6	1,116.2	161.3	10.2	3.4	2.5
West Virginia	2,507.0	364.5	592.0	475.9	608.9	465.7

Source: Council on Environmental Quality, *Coal—Surface Mining and Reclamation*, p. 4.

Table 30. Sulfur Content of Estimated Coal Reserves in Four Appalachian Coal Basin States
(in percent)

	Low 0 to 1.0%	Middle 1.1 to 2.0%	High 2.1+%
Eastern Kentucky	75.2	13.4	11.3
Ohio	1.3	5.9	92.6
Pennsylvania	2.1	34.6	63.3
West Virginia	46.1	34.2	19.7

Source: Bureau of Mines, *Information Circular* 8312, referenced in Charles River Associates, *Economic Impact of Public Policy*, p. 264.

regions in the Appalachian Coal Basin. They show that utilities desirous of complying with the impending EPA clean-air standards will, in the absence of sulfur-removal equipment, have to purchase coal from eastern Kentucky or West Virginia. This is particularly true for public utilities located in the eastern and southeastern United States because transporting western low-sulfur coal to these markets is prohibitively expensive. When the estimated life of remaining low-sulfur reserves is considered in terms of years (see Table 32), it is evident that only eastern Kentucky and West Virginia can hope to supply the demand for this type of coal for any length of time. As the table shows, at 1971 production rates, eastern Kentucky's reserves would last roughly 119 years, West Virginia's, 146 years. Ohio has practically no low-

Table 31. Recoverable Surface Mining Reserves by Estimated Sulfur Content in Four Appalachian Coal Basin States, 1974
(million short tons)

	Recoverable Surface Mining Reserves[a]	Sulfur Level		
		Low 0 to 1.0%[b]	Middle 1.1 to 2.0%[b]	High 2.1+ %[b]
Eastern Kentucky	3,105	2,111.4	751.4	239.1
West Virginia	4,691	2,514.4	1,487.0	690.0
Pennsylvania	1,063	0.0	317.8	745.2
Ohio	3,289	0.0	401.3	2,887.7

[a]Represents 90 percent of demonstrated surface coal reserves.

[b]May not add up to total due to rounding.

Source: Bureau of Mines, "Demonstrated Coal Reserve Base of the United States on January 1, 1974," *Mineral Industry Surveys,* 1974.

sulfur coal left, and reserves in Pennsylvania would last only a generation.

The four regions, however, are endowed much more generously with medium- and high-sulfur coal. The estimated life of these two classes of coal ranges from a low of seventy years for medium-grade eastern Kentucky coal to 600 years for the same type of coal deposits in Ohio. The statistics also show that the life of medium-sulfur reserves is nearly twice that of high-sulfur coal, at least insofar as 1971 production statistics are concerned. A shift in demand for more low-sulfur coal would, of course, accelerate the depletion of these reserves while at the same time, with output remaining constant, it would extend the life of the medium- and/or high-sulfur reserves. The life of the low-sulfur reserves, in this case, could be shortened significantly. With the current focus on burning clean coal, this is in fact what is beginning to happen.

A recently completed study shows that conventional coal-reserve estimates are based on the simple addition of coal tonnage without regard to heat content.[27] If the latter is taken into account, as is common practice with electric utilities who are the major consumers of coal, then present estimates of low-sulfur coal reserves may be overstated.[28] At the same time, high-sulfur coal abounds.

THE SIZE OF COAL-MINING FIRMS

In eastern Kentucky, just as in the Appalachian Coal Basin and the nation as a whole, coal is sold either on long-term contract or in the short-term marketplace, usually referred to as the spot market. Many steel producers, elec-

Table 32. Estimated Life of 1974 Coal Reserves by Sulfur Content Based upon 1971 Production Rates for Four Appalachian Coal Basin States
(in years)

	Low 0 to 1.0%	Medium 1.1 to 2%	High 2.1+%	Total
Eastern Kentucky	118.9	69.8	130.7	109.7
Ohio	—	599.8	231.9	244.4
Pennsylvania	29.0	115.7	308.1	173.5
West Virginia	145.6	554.7	120.1	184.2

Note: 1971 is the latest year for which production data by sulfur content are available from the Bureau of Mines. 1974 reserves shown in Table 27 have been reduced by 50 percent for underground mining, and by 10 percent for surface mining.

Sources: Calculated from Tables 27 and 30 and unpublished Bureau of Mines production figures for 1971, classified by sulfur content.

tric utilities, and other coal users are vertically integrated, that is, possess their own mining operations. Output from such captive mines is generally used internally and does not enter the market in significant amounts and therefore does not influence market price directly.

By far the largest share of coal output from the major regions in the Appalachian Coal Basin is sold commercially in the open market. This market is either a contract market or a spot market. An estimated 80 percent of all coal marketed is sold in the contract rather than in the spot market, because large users of coal, such as electric utilities, derive considerable benefit from being able to operate with the knowledge of having a secure source of coal supply. Mining operators likewise prefer to have secure sales contracts, because the longer the sales contract, the easier it is for them to obtain needed investment capital for expansion. The amount of coal sold on the spot markets is therefore rather small, and prices are highly sensitive to even moderate changes in the demand for and supply of coal.

Table 33 shows the relative amounts of coal sold commercially either on contract or in the spot market and by captive mines. Remarkable is the fact that with the exception of the 1972 Ohio output, the shares sold commercially and by captives were almost identical for West Virginia, eastern Kentucky, and Ohio for the years examined. Only in Pennsylvania do captive mines produce a much larger share of the output, both relatively and absolutely than in the three other subregions of the basin. The commercial market for coal in Pennsylvania is considerably smaller than in the other three

Table 33. Coal Sold Commercially and by Captive Mines in Four Appalachian Coal Basin States, 1969-1973
(million short tons)

	Coal Sold Commercially	Percent of Total	Coal Sold by Captive Mines	Percent of Total
1969				
West Virginia	121.2	86	19.8	14
Eastern Kentucky	54.1	88	7.5	12
Pennsylvania	49.7	63	28.9	37
Ohio	44.9	88	6.4	12
1970				
West Virginia	125.2	87	18.8	13
Eastern Kentucky	64.1	88	8.4	12
Pennsylvania	51.0	63	29.5	37
Ohio	48.8	88	6.6	12
1971				
West Virginia	105.1	89	13.2	11
Eastern Kentucky	64.0	89	7.6	11
Pennsylvania	47.7	65	25.2	35
Ohio	47.0	91	4.5	9
1972				
West Virginia	116.5	94	7.2	6
Eastern Kentucky	63.4	95	5.5	5
Pennsylvania	56.7	75	19.3	25
Ohio	50.1	98	0.8	2
1973				
West Virginia	103.2	89	12.2	11
Eastern Kentucky	65.6	89	8.4	11
Pennsylvania	52.9	69	23.5	31
Ohio	40.8	89	5.0	11

Note: No distinction is made between coal sold on contract and in the open market.
Sources: Bureau of Mines, *Minerals Yearbook*, 1969-1974; for eastern Kentucky, unpublished data.

states. An explanation for this may be the fact that many large steel plants whose parent companies own captive coal mines are located in Pennsylvania. This is not the case for eastern Kentucky or West Virginia.

No published data exist on the amount of coal sold on contract or in the spot market for individual states or their subregions. It is, however, possible to estimate that in 1973 only approximately 46 million short tons of coal entered the spot market in the four major coal-producing regions of the Appalachian Coal Basin.[29] This represents only approximately 15 percent of

Table 34. The Largest Eastern Kentucky Coal-Mining Firms, 1973 Production

		Output Short Tons	Percent of Eastern Kentucky Total	Cumulative Percent
1	U.S. Steel	4,174,778	5.6	5.6
2	J. L. Jackson (Breathitt Coal Co. & Others)	3,625,542	4.9	10.5
3	W. L. Meadows (Eastern Coal Corp. & Others)	2,246,229	3.0	13.5
4	Beth-Elkhorn	2,225,605	3.0	16.5
5	J. W. Hoffman (Scotia & Blue Diamond)	1,807,932	2.4	18.9
6	Mountain Drive Coal Co.	1,670,000	2.3	21.2
7	J. A. Sigmon (Coal Reserves Corp. & Others)	1,548,833	2.1	23.3
8	Eastover Mining Co.	1,480,401	2.0	25.3
9	National Mines Corp.	1,417,321	1.9	27.2
10	Elmer Whitaker (River Processing & Others)	1,327,700	1.8	29.0
11	Shamrock Coal Co.	1,270,798	1.7	30.7
12	Martin Co. Coal Corp.	1,119,089	1.5	32.2
13	Kentucky Carbon Corp.	1,143,465	1.5	33.7
14	Wolf Creek Collieries Co.	1,139,915	1.5	35.2
15	Pikeville Coal Co.	932,561	1.3	36.5
16	Marty Corp.	859,094	1.2	37.7
17	Diamond Coal Co.	854,903	1.2	38.9
18	Republic Steel Corp.	803,900	1.1	40.0
19	West Virginia Rebel Co.	800,073	1.1	41.1
20	J. P. Baugues (Harlan No. 4 Coal Co. & Others)	734,612	1.0	42.1
	Sub-Total 1-20	31,182,751		42.1
	Sub-Total 21-30	6,517,825	8.8	
	Sub-Total 1-30	37,700,576		50.9
	Sub-Total 31-40	4,653,696	6.3	
	TOTAL 1-40	42,354,272		57.2

Note: Total output for Eastern Kentucky in 1973 was 73,943,520 short tons.
Source: Kentucky Department of Mines and Minerals, *Annual Report*, 1973, pp. 22-79.

total output from these four states. With this relatively small amount of coal offered in the spot market, it is not surprising that a sharp increase in short-term demand, as occurred in 1974, would result in a vigorous increase in the spot-market price of coal.[30] Rising prices do, of course, create an incentive to increase output. This is accomplished either by expanding output in existing mines, to the extent feasible, or by resuming operations in heretofore unprofitable mining firms, now that higher prices increase the likelihood of profits. Sustained higher prices, however, will also encourage the entry into the industry of new firms who have never before mined coal. An example of this is the recent entry of firms previously active in heavy construction. Their entry, however, is confined largely to surface mining.

An examination of the larger firms operating in eastern Kentucky reveals that in terms of output, and disregarding firm ownership, market concentration in eastern Kentucky is small. Table 34 shows that no single mining company produced in 1973 more than 5.6 percent of eastern Kentucky's output and that the twenty largest mining companies together produced only 42 percent of total output. Not until the output from the thirty largest firms is aggregated is the 50 percent of output threshold exceeded. Adding the next ten largest firms increases the cumulative total for the region by only 6 percent. In other words, coal production in eastern Kentucky is widely distributed over a large number of firms. This distribution supports the commonly held view that coal producers in eastern Kentucky have a tendency to behave competitively because no single firm or small group of firms controls a sufficiently large share of production to be able to influence price to its own benefit.

From the firm output data in Table 34, it is evident that most of the typical coal firms operate more than one mine in the state. A list of the thirty-three largest coal mines, in descending order of size, is shown in Table 35. Mines are classified not only in accordance with production, ownership class, and coal district but also by number of workers employed and the number of days the mine was operated.[31] The data show that an overwhelming number of these larger mining operations (mines that produce 425,000 tons per year or more) are of the underground type. Only seven of the thirty-three largest mines in eastern Kentucky, or 21 percent, were surface, strip, or auger operations, and only one of these has an annual output large enough to rank it within the twelve largest coal mines in eastern Kentucky. The inference that can be drawn from Table 35 is that most large mines in eastern Kentucky are of the underground type.

The data in the table unfortunately do not permit the calculation of productivity ratios for the different mines because no information is collected by state authorities on how many days of the year each worker listed on company rolls actually worked. In the absence of this information, no judgments can be made on relative mine productivities.

Coal output ranked by county is shown in Table 36. The table also shows the distribution of output between underground and surface mining. By far the largest coal-producing county in eastern Kentucky is Pike County, producing 25 percent of total output for that portion of the state and more than one-third of the underground coal mined in all eastern Kentucky. More than one-half of the underground coal produced in eastern Kentucky is taken

Table 35. Thirty-three Largest Coal Mines in Eastern Kentucky, 1973

Kentucky Rank	Owner Category	Mine Name	Mine Company	Controlling Company	District	Type	Kentucky State 1973 Production Short Tons	Number of Workers	Number Days Worked
1	Steel	Winifrede	U.S. Steel	U.S. Steel	Harlan	URC	1,750,492	207	243
2	Coal	Mountain Drive	Mountain Drive	Mountain Drive	Harlan	SRC	1,670,000	210	286
3	Coal	No. 18	Shamrock Coal	Jewell Coal and Coke	Hazard	UTC	1,270,798	165	258
4	Coal	Kentucky Carbon	Kentucky Carbon	Carbon Fuel	Pikeville	URC	1,143,464	324	239
5	Coal	Kencar 1	Kentucky Carbon Corp.	Carbon Fuel Co.	Pikeville	URC	1,143,436	324	239
6	Coal	File No. 10219	Martin County Coal Co.	A.T. Massey Coal	Martin	USC	1,119,089	391	230
7	Steel	Chisholm	Pikeville Coal Co. Inc.	Steel Co. of Canada	Pikeville	URC	932,561	216	238
8	Steel	No. 26	Bethlehem-Elkhorn Corp.	Bethlehem Steel	Hazard	URC	903,410	340	253
9	Steel	No. 22	Bethlehem-Elkhorn Corp.	Bethlehem Steel	Hazard	URC	807,220	304	265
10	Steel	So. Winifrede	U.S. Steel	U.S. Steel	Harlan	URC	805,547	145	244
11	Steel	Lynch No. 32	U.S. Steel	U.S. Steel	Harlan	URC	772,621	280	242
12	Coal	No. 3	Wolf Creek Collieries Co.	A.T. Massey	Martin	URC	770,897	157	234
13	Steel	Republic	Republic Steel Corp.	Republic Steel Corp.	Pikeville	URC	764,345	220	246
14	Coal	File No. 8977	Marty Corp.	Marty Corp.	Martin	SAC	745,837	155	169
15	Other	Stone	Eastern Coal Corp.	Pittston Co.	Pikeville	URC	714,255	571	250
16	Coal	No. 1	Peter Cave Coal Co. Inc.	Herman Mills	Martin	UTC	692,117	184	241
17	Coal	No. 5A	Diamond Coal Co.		Martin	ASTC	606,152	38	155
18	Other	Leatherwood No. 1	Blue Diamond Coal	W.R. Grace Co.	Hazard	URC	590,344	369	242
19	Steel	Stinson No. 1	National Mines	National Steel	Martin	URC	542,771	159	237
20	Other	Scotia	Scotian Coal Co.	W.R. Grace Co.	Hazard	URC	535,987	259	243
21	Coal	No. 76	Breathitt Coal Corp.	J.L. Jackson Falcon Coal Co.	Hazard	ASTC	531,280	54	225
22	Utility	Darby 4	Eastover Mining Co.	Duke Power Co.	Harlan	URC	530,783	194	228
23	Other	No. 2	Johns Creek Elkhorn Coal Corp.	General Energy Corp.	Pikeville	UTC	526,341	102	251

24	Coal	Coal Resources Corp.	Belmon	Ray Resources Corp.	Harlan	UTC	492,938	50	245
25	Utility	Eastover Mining Co.	Baker 1	Duke Power Co.	Harlan	UTC	491,493	85	247
26	Other	Kentland Elkhorn Coal Corp.	No. 2	Pittston Co.	Pikeville	URC	484,126	212	240
27	Coal	Breathitt Coal Corp.	No. 84	J.L. Jackson Falcon Coal Co.	Hazard	ASTC	479,700	50	270
28	Coal	Ryan's Creek Coal Co. Inc.	No. 3	Logan & Kanawha Coal Co. Inc.	Harlan	UTC	462,094	88	234
29	Coal	Western Coal Corp.	No. 1	J.L. Jackson Falcon Coal Co.	Martin	STC	451,146	24	209
30	Coal	Breathitt Coal Corp.	No. 77		Hazard	ASTC	446,500	61	180
31	Coal	South-East Coal Co.	Polly 4	South-East Coal Sales Co.	Hazard	URC	426,675	236	256
32	Coal	Martin County Coal Corp.	No. 1C	A.T. Massey (Sales)	Martin	URC	426,011	103	244
33	Coal	Kentucky Central Coal Corp.	No. 1	B. W. McDonald	Martin	ASTC	425,014	26	209

Sources: Kentucky Department of Mines and Minerals, *Annual Report*, 1973; *Keystone Coal Industry Manual*, 1974.

Key: S   Surface Mine   T   Truck transport
     U   Deep mine   R   Rail transport
     A   Auger mine   C   Coal mine

Table 36. Eastern Kentucky Coal Production, 1974[a]

County[b]	Underground		Surface		Total County Production	
	Tonnage	% of Eastern Kentucky Total	Tonnage	% of Eastern Kentucky Total	Tonnage	% of Eastern Kentucky Total
Pike**	14,447,794	35.048	7,220,506	16.49	21,668,300	25.487
Harlan	8,196,074	19.882	2,021,059	4.62	10,217,133	12.018
Martin	2,332,970	5.659	4,220,618	9.64	6,553,588	7.708
Perry	2,128,315	5.163	3,409,337	7.79	5,537,652	6.513
Letcher	3,570,958	8.663	1,900,614	4.33	5,471,572	6.436
Floyd	2,812,476	6.823	2,480,371	5.66	5,292,847	6.226
Breathitt*	56,291	.137	5,053,359	11.54	5,109,650	6.010
Knott*	3,319,367	8.052	1,253,836	2.86	4,573,203	5.379
Bell	1,048,799	2.544	3,276,819	7.48	4,325,618	5.088
Leslie	1,674,673	4.063	1,376,058	3.14	3,050,731	3.588
Whitley	216,369	.525	1,519,642	3.47	1,736,011	2.042
Knox	39,158	.095	1,459,576	3.33	1,498,734	1.763
Johnson	140,233	.340	1,265,549	2.89	1,405,782	1.653
Magoffin	—	—	1,343,418	3.07	1,343,418	1.580
Lawrence	11,792	.029	1,176,592	2.69	1,188,384	1.398
Laurel	76,269	.185	951,185	2.17	1,027,454	1.209
McCreary	674,915	1.637	150,037	.34	824,952	.970
Morgan	—	—	745,323	1.70	745,323	.877
Clay	315,929	.766	374,234	.86	690,163	.812
Elliott	—	—	678,152	1.55	678,152	.798
Carter	2,489	.006	548,212	1.25	550,701	.648
Pulaski	127,602	.309	326,362	.75	453,964	.534
Owsley	—	—	317,759	.73	317,759	.374

Lee	20,300	.049	225,109	.51	245,409	.289
Boyd	—	—	184,964	.42	184,964	.218
Greenup	—	—	179,180	.41	179,180	.211
Jackson	500	.001	53,250	.12	53,750	.063
Wayne	5,764	.014	40,726	.09	46,490	.055
Rockcastle	1,518	.004	30,032	.07	31,550	.037
Wolfe	—	—	13,942	.03	13,942	.016
Menifee	1,307	.003	—	—	1,307	.002
Totals[c]	**41,221,862**	**99.997**	**43,795,821**	**99.59**	**85,018,683**	**99.993**

[a]Ranked by total production.
[b]Divided counties: *Hazard-Martin **Pikeville-Hazard
[c]Data do not add to 100 because of independent rounding.

Source: Calculated from Kentucky Department of Mines and Minerals, *Annual Report*, 1974.

from Pike and Harlan counties. Other major underground mining counties are Letcher, Knott, and Floyd.

While Pike County is the largest surface-mining county in the region, Martin and Breathitt also contribute a sizable portion of the output. Surface mining is much more evenly distributed over the thirty-one eastern Kentucky coal counties than underground mining. With eight exceptions, surface mining generates more output than deep mining in the eastern Kentucky counties, and there are several counties in which there exists surface mining to the exclusion of deep mining.

Map 5 shows the coal-producing counties of Kentucky. From this map, the concentration of coal production in the southeastern and eastern portions of eastern Kentucky are discernible. This is the most mountainous area of the state. Further west there is a declining rate of coal output with fringe counties such as Wolfe, Menifee, Greenup, Owsley, and Elliott representing recent entrants into the community of coal producers. As the demand for coal and its price escalated in recent years, so has the activity of coal operators in search of mineral lands. The spillover into counties lying on the western fringes of Appalachia is a manifestation of this trend.

Several of the traditionally largest coal-producing counties in eastern Kentucky have shown a relatively small or, in some cases, a declining rate of growth in coal output in the past five years. Pike, Letcher, Perry, and Floyd are some of these counties (see Table 37). Harlan County, on the other hand, the second-largest coal-producing county in eastern Kentucky shows 27.5 percent increase in coal output over the 1969 to 1974 period.

Underground mining is the dominant form of mining in at least four of these five counties, and expansion of this type of mining typically takes place only over the longer term, and then only if coal operators perceive of a relatively stable future. Implementation of the Coal Mine Health and Safety Act of 1969, however, created considerable uncertainty in the industry and contributed greatly, at least in 1970 and 1971, to a reluctance to expand underground operations. As a consequence the bulk of the additional demand for coal in 1973 and 1974 (and probably for a few more years in the future) was satisfied from surface operations, wherever these could be expanded, either in traditional coal counties or in counties where little coal had been surface-mined previously. In fact, the rate of growth of coal output in some counties such as Martin, Whitley, Morgan, Boyd, Laurel, Carter, Knox, Lee, Jackson, and Rockcastle has been nothing short of dramatic in the past few years. Several new counties joined the list of coal producers, where coal is being mined only by surface methods. These counties are Elliott, Owsley, Greenup, and Wolfe.

It is clear, therefore, that a sharply rising demand for coal is being satisfied largely from new surface mining in eastern Kentucky. Underground mining

Map 5. Kentucky Coal Output by County, 1974

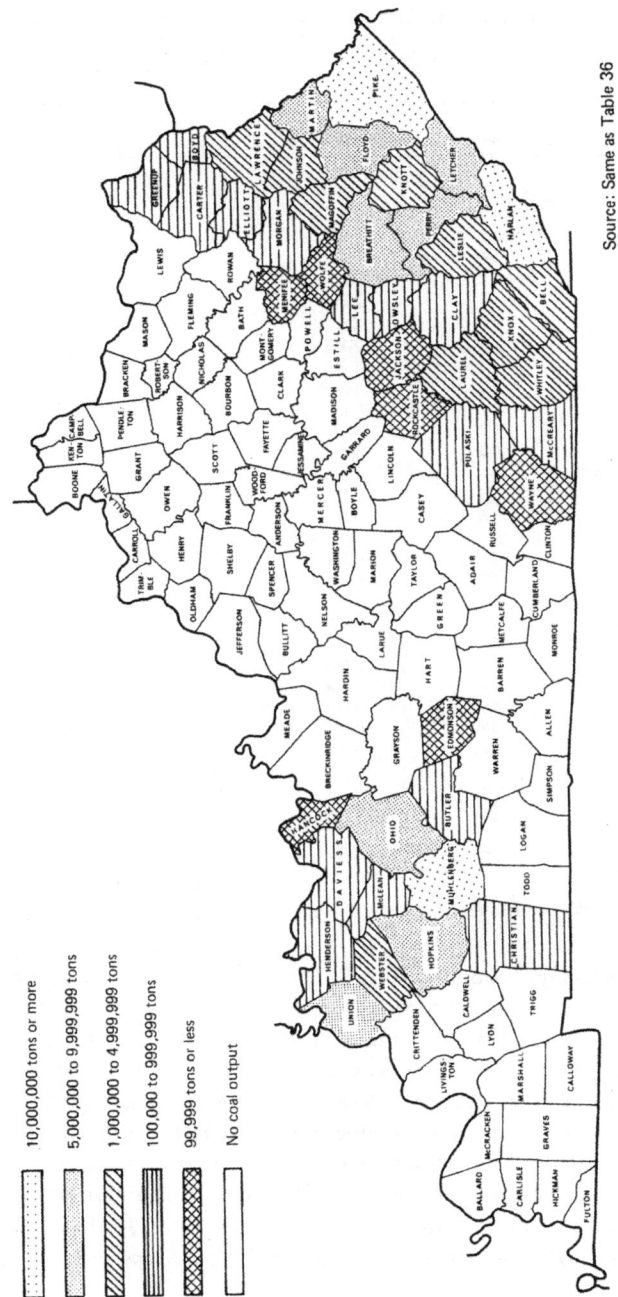

10,000,000 tons or more
5,000,000 to 9,999,999 tons
1,000,000 to 4,999,999 tons
100,000 to 999,999 tons
99,999 tons or less
No coal output

Source: Same as Table 36

Table 37. Growth in Coal Output
by Eastern Kentucky Counties, 1969-1974
(million short tons)

County	Rank	1969	Rank	1974	Percent Change
Pike	1	21.3	1	21.7	1.87
Harlan	2	8.0	2	10.2	27.50
Letcher	3	6.2	5	5.5	−11.30
Perry	4	6.1	4	5.5	− 9.84
Floyd	5	4.5	6	5.3	17.78
Knott	6	2.7	8	4.6	70.37
Breathitt	7	2.6	7	5.1	96.15
Bell	8	2.2	9	4.3	95.45
Leslie	9	1.8	10	3.1	72.22
Martin	10	1.3	3	6.6	407.69
Johnson	11	0.8	13	1.4	75.00
Magoffin	12	0.6	14	1.3	116.66
Pulaski	13	0.6	22	0.5	−16.67
Clay	14	0.5	20	0.7	40.00
McCreary	15	0.5	17	0.8	60.00
Whitley	16	0.3	11	1.7	466.67
Lawrence	17	0.3	15	1.2	300.00
Knox	18	0.2	12	1.5	650.00
Wayne	19	0.07	28	0.05	−28.57
Morgan	20	0.03	18	0.7	2,233.33
Lee	21	0.02	24	0.2	1,000.00
Clinton	22	0.009	–	–	–
Boyd	23	0.007	26	0.2	2,757.14
Laurel	24	0.006	16	1.0	16,566.67
Carter	25	0.005	21	0.6	11,900.00
Jackson	26	0.003	27	0.05	1,566.67
Rockcastle	27	0.002	29	0.03	1,400.00
Elliott	28	–	19	0.7	–
Owsley	29	–	23	0.3	–
Greenup	30	–	25	0.2	–
Wolfe	31	–	30	0.01	–
Menifee	32	–	31	0.001	–

Sources: Kentucky Department of Mines and Minerals, *Annual Report,* 1970, 1975.

continues to supply about one-half of the output, usually by well-established operators with long-term contracts.

In 1973 more than 3,400 mines produced 307,000 tons of coal in eastern Kentucky, West Virginia, Pennsylvania, and Ohio. The average mine differs from region to region, its size determined largely by geological formations, state regulations concerning mining and operating procedures, size of market and access to it, and other factors. In order to appreciate fully the differences in size of mining operations in the four regions, it is important to develop a size distribution of coal mines for each region. Table 38 shows such a distribution for 1973. Mines are classified into six groups in accordance with

Table 38. Size Distribution of Mines among Four Appalachian Coal Basin States, 1973
(million tons)

	Category 1		Category 2		Category 3		Category 4		Category 5		Category 6		Total	
	No. of Mines	Output	No. of Mines	Output	No. of Mines	Output	No. of Mines	Output	No. of Mines	Output	No. of Mines	Output	No. of Mines	Output
Eastern Kentucky														
Underground	12	10.0	27	9.2	49	7.0	63	4.2	282	6.5	237	1.1	670	38.0
Surface*	4	2.3	20	6.8	55	7.5	97	6.6	292	6.1	220	1.6	688	30.9
													1,358	68.9
Ohio														
Underground	14	12.7	8	2.9	2	0.4	2	0.2	5	0.1	4	.02	35	16.3
Surface*	17	18.7	23	6.8	28	3.6	46	3.2	88	2.1	69	0.4	271	34.7
													306	51.0
Pennsylvania														
Underground	40	35.1	27	9.7	20	2.7	12	0.9	23	0.5	37	0.1	159	49.1
Surface*	1	0.7	10	2.6	50	7.0	83	5.8	362	9.8	171	0.9	677	26.8
													836	75.9
West Virginia														
Underground	49	49.7	92	30.1	61	9.5	70	5.7	161	5.7	115	0.9	548	101.7
Surface*	4	3.1	18	4.6	45	5.8	72	5.4	154	2.7	94	0.5	387	22.1
													935	123.8

*Surface includes both strip and auger mines.

Categories: 1 – over 500,000 tons   3 – 100,000 - 200,000   5 – 10,000 - 50,000
            2 – 200,000 - 500,000   4 – 50,000 - 100,000   6 – Under 10,000

Source: Bureau of Mines, "Coal—Bituminous and Lignite in 1973," *Mineral Industry Surveys*, 1975.

Table 39. Average Mine Output for Four
Appalachian Coal Basin States, 1973

	Output Per Mine in Short Tons	
	Underground	Surface
Eastern Kentucky	56,637	44,929
Ohio	464,829	128,037
Pennsylvania	309,013	39,595
West Virginia	185,515	57,054

Source: Calculated from Table 38.

annual output and type of mining. The largest number of mines, although not those with the largest outputs, was found in eastern Kentucky where 1,362 mines produced 73.9 million tons of coal in 1973. The smallest number of mines, 235, was found in Ohio, even though total output for that year exceeded 45 million tons.

In terms of average mine size, measured in tons produced annually, the smallness of the typical eastern Kentucky mine stands out. Table 39 shows that the average eastern Kentucky underground mine produces far less coal than competitive mines in Ohio, Pennsylvania, and West Virginia. Only surface mines in Pennsylvania seem to be somewhat smaller than eastern Kentucky mines, but not by very much.

The remarkable fact that emerges from Table 39 is that the average size of the eastern Kentucky underground mine, measured in annual tons of coal produced is approximately one-third to one-eighth, respectively, the size of similar mines located in the adjoining states.[32] And while there are in operation some very large mines in eastern Kentucky (see Table 34), the preponderance of mines is small. More than one-third of the underground mines operating in the region produced fewer than 10,000 tons in 1973. This fact is substantiated further by the percentage distributions shown in Table 40. In eastern Kentucky, it took three-fourths of all underground mines (i.e. the smaller mines of categories 5 and 6) to produce but 17 percent of total underground-coal output, while in two other regions, roughly the same percentage of mines produced considerably more coal: 22 percent of West Virginia underground output and approximately 26 percent of Pennsylvania underground output.

Conversely, 40 percent of the underground mines in Ohio were in the 500,000-tons-or-larger category and produced 94 percent of that state's underground output, while in Pennsylvania only 29 percent of the underground mines fell into this class and produced 74 percent of the state's output.

Table 40. Percentage Distribution of Mines and Output by Size for Four Appalachian Coal Basin States, 1973

	Category 1		Category 2		Category 3		Category 4		Category 5		Category 6	
	Percent of Mines	Percent of Output	Percent of Mines	Percent of Output	Percent of Mines	Percent of Output	Percent of Mines	Percent of Output	Percent of Mines	Percent of Output	Percent of Mines	Percent of Output
Eastern Kentucky												
Underground	1.8	26.4	4.0	24.2	7.3	18.3	9.4	11.1	42.1	17.1	35.4	2.9
Surface	0.6	7.3	2.9	22.1	8.0	24.3	14.1	21.4	42.4	20.0	32.0	5.1
Ohio												
Underground	40.0	78.3	22.9	17.5	5.7	2.3	5.7	1.0	14.3	0.7	11.4	0.1
Surface	6.3	54.0	8.5	19.6	10.3	10.2	17.0	9.1	32.5	6.0	25.5	1.0
Pennsylvania												
Underground	25.2	71.5	17.0	19.8	12.6	5.6	7.5	1.8	14.5	1.1	23.3	0.2
Surface	0.1	2.5	1.5	9.8	7.4	26.2	12.3	21.6	53.5	36.5	25.3	3.4
West Virginia												
Underground	8.9	48.9	16.8	29.6	11.1	9.4	12.8	5.6	29.4	5.6	21.0	0.9
Surface	1.0	14.1	4.7	21.0	11.6	26.3	18.6	24.3	40.0	12.1	24.3	2.2

Note: Figures may not sum due to rounding.

Source: Calculated from Table 38.

Figure 2. Relationship between the Percentage of Underground-Mining Operations and the Percentage of Coal Produced, 1973

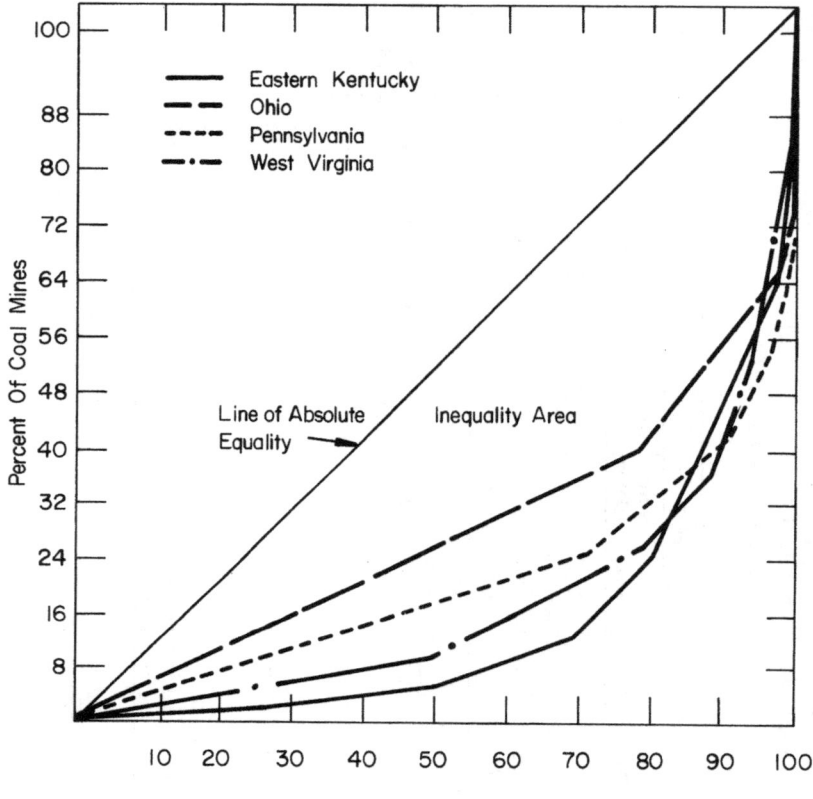

Source: Table 40.

A graphic representation of the cumulative relationship between percentage of underground coal mines and percentage of output is shown in Figure 2. A 45-degree line relationship would denote an equal distribution, that is, a given percentage of mines would contribute an equal percentage of total output. An equal distribution would exist, for example, if 8.9 percent of eastern Kentucky's mines produced 8.9 percent of the output, or if 34 percent of the mines in Ohio produced 34 percent of the output. The more convex (from above) the curve which represents the relationship between percentage of firms and percentage of output, the greater the inequality between the two measures, that is, the greater the share of total underground output accounted for by a small percentage of mines. Figure 2 clearly shows that for all four regions a small percentage of mines accounts for a disproportionate share of output. This is less the case for West Virginia and least for Ohio. For eastern Kentucky, the inequality is greatest, where 8.9 percent of mines accounted for 60 percent of the 1973 output, while in Ohio, approximately 34 percent of the mines accounted for that same percentage of output. In short, Figure 2 shows that in underground mining, a relatively small number of firms contributes the major portion of eastern Kentucky coal output, while the large number of small mines contributes only a small amount. In West Virginia, Pennsylvania, and Ohio, underground output is accounted for by relatively more mines. The likelihood that market dominance by a few mines could develop is therefore less in these three regions than in eastern Kentucky, even though in the latter a very large number of mines are in operation. But most of these are very small and therefore are price-takers having no significant influence in the market for coal.

In addition to the size distribution of underground mines, Table 38 also shows the distribution for surface mines. Given the geological formations of Pennsylvania, West Virginia, and eastern Kentucky, it is not surprising that few mines fall into the 500,000-tons-per-year output range in these regions. In Ohio, on the other hand, where the topography of the land is much flatter, mines of that size are not uncommon. Most of the mining operations in Ohio are able to use draglines for stripping, whereas in the other three regions such large equipment would not be very useful. Mobility is restricted by the narrow hollows and steep mountainsides which characterize the land.

The smaller average size of surface mines in eastern Kentucky, Pennsylvania, and West Virginia as compared to Ohio is also shown in Table 39. Ohio's average mine is from three to four times larger than those in the neighboring states.

If surface mines of 50,000 tons per year of output or smaller are considered, that is, those that fall into the two smallest size categories, then, as Table 40 shows, in eastern Kentucky these mines produced 25 percent of surface output, in West Virginia 14 percent, and in Pennsylvania 40 percent. In Ohio, on the other hand, mines of this size produced only 7 percent of the state's output from surface mines even though they make up 58 percent of the total number of surface mines in operation. It becomes clear once again that in Ohio a relatively small percentage of surface mines (14 percent of the largest surface mines in the state) accounted for nearly three-fourths of the output. If categories 1 and 2 are examined for eastern Kentucky and West Virginia, it is evident that for these two regions large surface mines also account for a disproportionate share of state output, 29 and 35 percent, respectively. Thus while surface operations are large in these two regions, they are not quite as important in terms of total output as mines of comparable size in Ohio.

Only in Pennsylvania are the large mines relatively less important. They constitute 1.6 percent of the total surface-mine population and contribute 12.3 percent of surface output.

Figure 3 presents a pictorial view of mine-size distribution related to share in output. The curves show that Ohio has the least equal distribution in percentage of mines contributing to total surface output. Fourteen percent of the mines contributed more than two-thirds of surface output, while in eastern Kentucky 10 percent of the mines produced 53 percent of output. West Virginia's distribution is very similar to that of eastern Kentucky, while the number of Pennsylvania mines is more consistent with their share in output, that is, they are more evenly distributed.

In comparing size distribution of mines in underground mining with that in surface mining in the four area regions (see Figures 2 and 3) it is clear that differences among the four subregions are more pronounced in the former than in the latter. That is, the sizes of mines in underground mining differ more among subregions than in surface mining. Eastern Kentucky appears to have the smallest percentage of large underground mines, Ohio the largest. Pennsylvania has the smallest percentage of large surface mines, Ohio the largest. In short, Ohio is a region where mining tends to be of larger capacity both in underground and in surface operations. Geological formations undoubtedly play a key role there in determining mine size.

In addition to an examination of mine-size distribution for 1973, we shall inquire into possible changes in this distribution that may have taken place over time. Table 41 shows the percentage share of output that was accounted for by various sizes of underground mines from 1969 to 1973. The statistics demonstrate that there has been little change in the share of output accounted for by the smallest three classes in the four regions.

Only the smallest class, 10,000 tons and less per year, shows a decline in its share of total output. This is true for eastern Kentucky, Ohio, and Pennsylvania, but not for West Virginia. These small mines find it unprofitable to install the requisite safety equipment which is now required under the new Federal Coal Mine Health and Safety Act and therefore do not have the incentive either to expand current output or to enter into new underground ventures. What is more, some small operators whose businesses were only marginally profitable ceased mining altogether. For them, the additional costs of installing the newly required safety equipment would have meant incurring losses.

Figure 3. Relationship between the Percentage of Surface-Mining Operations and the Percentage of Coal Produced, 1973

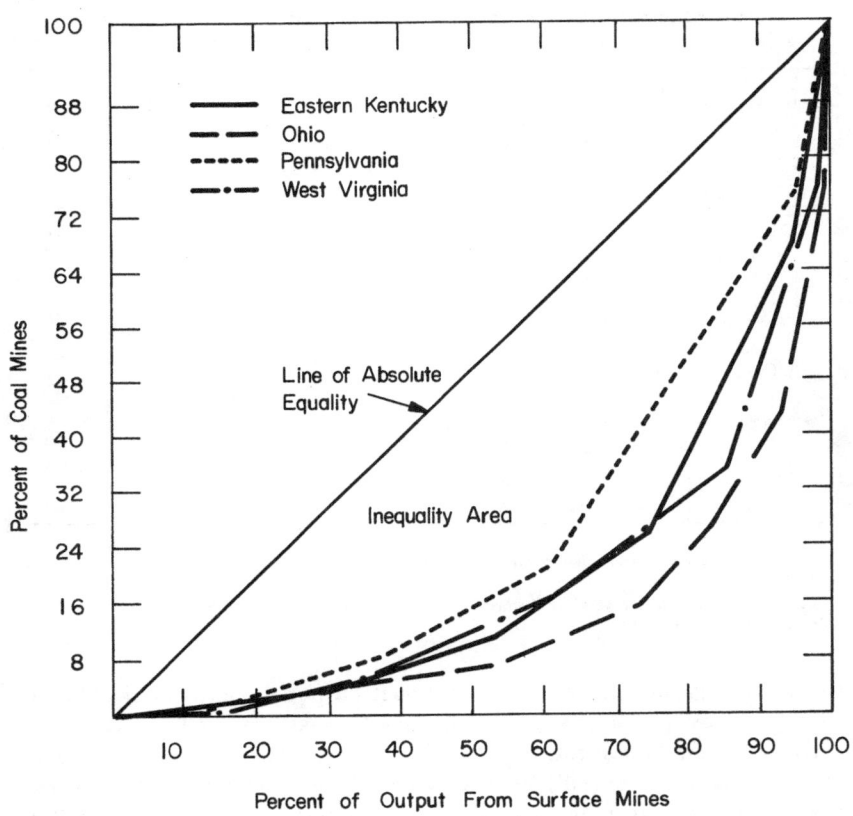

Source: Table 40.

The share of output accounted for by the largest class, mines with output in excess of 500,000 tons, remained about the same over the 1969-1973 period for the four regions although year-to-year fluctuations were often large. The sharpest changes occurred in eastern Kentucky and West Virginia where that class's output share fell from 1969 to 1973 by 10 percent.

In contrast, the shares of mines with an annual output of 100,000 to 500,000 tons generally increased or stayed the same in the four regions between 1969 and 1973.

The evidence depicted in Table 41 suggests that for the four regions exam-

Table 41. Percentage Share of Output from Different Size Underground Mines in Four Appalachian Coal Basin States, 1969-1973

		Category					
	Year	1	2	3	4	5	6
Eastern Kentucky	1969	39	17	11	11	17	6
	70	33	18	12	12	19	6
	71	29	20	12	15	19	6
	72	26	24	18	11	17	3
	73	34	26	14	10	14	2
Ohio	1969	93	4	0.9	1	0.9	0.3
	70	96	2	0	0.7	0.9	0.5
	71	83	12	2	1	2	0.2
	72	78	18	2	1	0.7	1
	73	94	3	1	1	0.6	0.1
Pennsylvania	1969	73	18	5	2	2	0.6
	70	76	14	7	2	1	0.4
	71	65	23	6	3	2	0.5
	72	72	20	6	2	1	0.2
	73	74	18	4	2	0.7	0.3
West Virginia	1969	57	27	7	4	4	1
	70	58	25	7	4	4	0.9
	71	50	29	10	6	5	0.8
	72	49	30	9	5	5	1
	73	51	27	12	6	4	0.5

Note: Percentage may not sum due to rounding. See Table 38 for size of categories.

Sources: Bureau of Mines, *Minerals Yearbook,* 1970-1973; Kentucky Department of Mines and Minerals, *Annual Report,* 1970-1973.

ined, underground mines with an annual output of between 50,000 to 500,000 tons are the most profitable ones. This inference is drawn from the fact that given the institutional environment which these mines faced during that period, they generally expanded their share in total production at the expense of other sizes of mines. It is therefore possible to argue that smaller or larger mines generate unacceptably high costs in the long run and that if one assumes that mines pursue the profit objective, they will gravitate over time toward that size which will allow them to produce at lowest cost. According to the available evidence, classes 2, 3, and 4 in Table 41 appear representative of that size.

For eastern Kentucky specifically, the data in Table 41 show that underground output is much more evenly distributed over the six size classes than in the other three subregions. In 1973 no single class accounted for more than 34 percent of the state's underground output, while in Ohio, 94 percent of output originated in the largest size class.

The shares of different sizes of surface mines in total output for the years

Table 42. Percentage Share of Output from Different Size Surface Mines in Four Appalachian Coal Basin States, 1969-1973

	Year	Category 1	2	3	4	5	6
Eastern Kentucky	1969	9	36	16	20	16	1
	70	7	31	18	20	15	8
	71	10	20	23	19	16	12
	72	7	22	24	21	20	5
	73	15	31	17	19	16	2
Ohio	1969	51	17	11	12	8	1
	70	58	17	9	10	6	0.8
	71	50	20	11	10	7	0.6
	72	54	20	10	9	6	1
	73	53	17	14	9	6	0.6
Pennsylvania	1969	3	5	29	25	34	4
	70	2	6	31	28	30	3
	71	15	14	24	24	20	3
	72	3	10	26	22	35	3
	73	4	4	21	25	43	3
West Virginia	1969	5	21	25	25	21	2
	70	9	18	22	23	26	2
	71	11	18	28	25	17	2
	72	14	21	26	24	12	2
	73	0	35	18	21	23	2

Sources: *Minerals Yearbook,* 1960 (vol. 1, Table 7), 1970 (vol. 1); "Coal—Bituminous and Lignite," *Mineral Industry Surveys,* preliminary release, 1971, 1972.

1969-1973 are shown in Table 42. In West Virginia the largest surface mines, those with an output in excess of 500,000 tons per year, no longer produce at that level of output. All other regions' mines in category 1 show little change in their share of total surface output over the five-year period. In West Virginia only size categories 2 and 5 seem to have expanded their share in total output. This implies that in this state these two classes may be regarded as most likely to yield long-term profits.

At the other end of the spectrum, mines in category 6, no change in output share has occurred in the regions. In category 5, on the other hand, eastern Kentucky and Pennsylvania mines have increased their share in total output. As the demand for low-sulfur coal began to surge in 1970, small surface operations began to emerge in large numbers, and their share in total output rose. And as the demand for this type of coal continued to grow unabated, many of these small mines began to expand operations so that by 1973 their annual output exceeded 10,000 tons. As a consequence the share of category 5 in total output in eastern Kentucky, Pennsylvania, and West Virginia rose significantly.

Expanding output or establishing new operations in the category of 10,000 to 50,000 tons per year is relatively easy in eastern Kentucky because the institutional environment, when compared with the neighboring states, favors surface mining.

Little change has occurred over the years in market shares of firms with an annual output of over 50,000 tons. Although differences in output in one year may throw a firm into categories different from the one usually occupied, as a whole a discernible growth trend, with the exception of Ohio, is observable only in category 5, mines producing from 10,000 to 50,000 tons per year. It is also true that large surface mines have not increased their share in total output, but rather have lost ground to medium-size mines.

The statistics shown in Table 42 also depict a fairly strong degree of stability in Ohio and Pennsylvania mine shares. The largest size surface mines continue to produce over one-half of total surface output, and there is little evidence to show that small operations can appropriate a substantial market share. In Ohio the explanation is found in the topography of the land which favors large mining operations. In Pennsylvania, and to some extent in West Virginia, on the other hand, it is the hilly terrain and the more vigorous enforcement of reclamation laws that seem to discourage the expansion of large surface operations. Until 1972, West Virginia's largest operators continued to prosper and increase their share in total output, but in 1973 output from this class of mines ceased completely.

Perhaps the greatest dissimilarity in surface mining among the four subregions is found in the large number of small mines in eastern Kentucky, where they accounted for more than one-third of the mines operative in 1973 (see Table 40). In West Virginia, Pennsylvania, and Ohio they accounted for between 18 and 22 percent of operating mines. Their contribution to total surface output, though disproportionately small, was still considerably greater in eastern Kentucky than in the other three regions. That the trend in the growth of small surface mines continued at least into 1974 is illustrated by the dramatic increase in surface-mine permits in that year, most of which were issued to small operators. Table 43 lists the growth in surface-mine licenses issued in the last five years in eastern Kentucky. In 1974 licenses almost doubled over 1973.

Many reasons exist to explain the rather sharp increase in small surface mines in eastern Kentucky. The explanations become more forceful when it is considered that in Pennsylvania and West Virginia, surface-mining permits issued actually declined in recent years. Studies have shown that in eastern Kentucky reclamation laws are not as stringently enforced as in neighboring states, due principally to the time lag associated with the rapid growth in surface mining and the need to provide for an adequate and appropriate in-

Table 43. Surface Mines Licensed in Eastern Kentucky, 1970-1974

Year	Number of Mines Licensed
1970	502
1971	740
1972	547
1973	646
1974	1,103

Sources: Kentucky Department of Mines and Minerals, *Annual Report,* 1969-1974.

spection system.[33] Initial bonding and permit costs are lower than in the other subregions of Appalachia.[34] Union membership and union restrictions are fewer in eastern Kentucky than in neighboring states. The legality of the broad form deed makes more potential coal reserves available than elsewhere where surface rights take precedent. The uneven enforcement of weight restrictions on eastern Kentucky roads makes transporting coal by truck less costly.[35]

The causes which led to the recent surge in small surface-mining activity in eastern Kentucky are undoubtedly a combination of all these points and not of any one factor exclusively.

PRODUCTIVITY IN COAL MINING

The concept of productivity refers to the measure of output per unit of input. The most widely used measure is output per man-hour. Output can also be expressed in terms of man-days, man-months, or even man-years. It must be remembered that productivity does not depend on labor effort alone but also on the kind and amount of machinery used, the type of raw materials available, and, in the coal industry, on geological and topographical configurations. It is also possible to measure output in terms of per unit of capital or per unit of machinery input, but the customary focus is on man-days because of the special emphasis placed on output per unit of human labor.

Up to approximately the mid-1960s, eastern Kentucky lagged behind other regions in mining productivity, mostly because the then-dominant forms of deep mining took place in relatively smaller mines that were not very capital-intensive. Today this is no longer the case, as eastern Kentucky mines are either more or at least as productive as mines located in neighboring states. Table 44 compares productivities in eastern Kentucky mining with those in Ohio, Pennsylvania, and West Virginia.

In underground mining alone, eastern Kentucky clearly ranks as the most

Table 44. Productivity in Coal Mining for Four Appalachian Coal Basin States, 1968-1973
(tons)

	Year	Underground		Strip	
		Output per Man-Day	Percentage Change	Output per Man-Day	Percentage Change
Eastern Kentucky	1968	14.81	+ 5.2	27.30	+43.3
	69	15.58	− 9.2	39.13	+12.3
	70	14.15	−12.2	43.93	−15.9
	71	12.42	− 0.4	36.94	− 8.0
	72	12.37		33.99	
	73	12.70	+ 2.7	34.75	+ 2.2
Ohio	1968	16.86	+ 2.8	33.75	+ 5.8
	69	17.34	−10.0	35.69	+ 8.3
	70	15.62	−27.7	38.65	+ 1.0
	71	11.29	+11.1	38.27	− 5.3
	72	12.54		36.24	
	73	11.89	− 5.2	36.88	+ 1.8
Pennsylvania	1968	14.91	− 6.7	19.56	+16.7
	69	13.92	−11.8	22.84	+ 1.8
	70	12.28	−14.1	23.26	+ 3.2
	71	10.56	− 3.5	24.00	+ 0.7
	72	10.19		24.18	
	73	9.63	− 5.5	23.11	− 4.4
West Virginia	1968	14.81	0.0	29.48	+ 6.2
	69	14.81	−15.6	31.31	−10.9
	70	12.50	−12.4	27.89	− 1.5
	71	10.96	+ 2.5	27.48	+11.7
	72	11.23		30.69	
	73	10.53	− 6.3	28.11	− 8.4

Note: With the exception of eastern Kentucky, very little coal is auger mined in the four Appalachian Coal Basin states; therefore, productivity differences in underground and strip mining only are considered here.

Sources: Bureau of Mines, *Mineral Industry Surveys,* 1968-1973.

productive of the four regions considered. The enactment of the Federal Coal Mine Health and Safety Act of 1969, as expected, had an adverse impact upon productivity in all four regions, but if eastern Kentucky data for 1971 to 1973 are taken as a trend, then it can be argued that the adverse effects have run their course. For Ohio, Pennsylvania, and West Virginia, the data do not reflect quite the same trend as that apparent for eastern Kentucky. That is, with the exception of 1972, productivity in those regions appears to be continuing its decline.

Comparing 1973 productivities in coal mining with those of 1968 clearly

Table 45. Productivity in Coal Mining for Four Appalachian Coal Basin States, 1968 and 1973

	Year	Underground Productivity per Man-Day	Strip Productivity per Man-Day
Eastern Kentucky	1968	1.000	1.000
	1973	.858	1.273
Ohio	1968	1.000	1.000
	1973	.705	1.093
Pennsylvania	1968	1.000	1.000
	1973	.646	1.181
West Virginia	1968	1.000	1.000
	1973	.711	.954

Source: Calculated from Table 44.

illustrates the declining trend in underground mining. Table 45 shows that in 1973 Pennsylvania mines were one-third less productive than in 1968. Eastern Kentucky mines, on the other hand, suffered only a 15 percent loss, less than one-half that of Pennsylvania.

There are two major factors operative which help to explain the nature of the decline in underground productivity. First, new safety legislation altered work procedures in underground mining so that a greater share of a miner's work-day has to be spent taking safety precautions. Therefore, correspondingly less time is devoted to the actual mining of coal. In addition, new personnel need to be employed to make up for the loss of work time devoted to safety procedures and maintain safe and healthy working conditions as specified by the law. All these factors clearly tend to reduce output per man-hour or man-day and thereby productivity.[36]

The second major change that occurred in the industry tended to mitigate, not enhance, the decline in underground productivity. Passage of the new federal safety legislation imposed upon mines additional costs in the form of newly required safety equipment. Many mines had only been marginally profitable because of relatively high production costs, whatever their origin, and were unable to meet the new labor and capital costs imposed upon them by the passage of the new legislation. They therefore left the industry. These were typically the smallest and least productive mines with the highest production costs. Their departure had the effect of pushing average productivity in the industry up to a level above that which would have obtained if these mines had remained in the industry.

In eastern Kentucky, more so than in the other three subregions, large numbers of small family-type mining operators with higher than average production costs were operative. When they ceased operating, the result was an upward movement in the average value of underground-mining productivity.

Figure 4 shows diagrammatically the change in mining productivity for underground and strip mining in the four Appalachian coal regions. The decline in underground productivity in the years following 1969 is clearly articulated.

In strip mining, the change in productivity over the past six years was much more moderate than in underground mining. Only eastern Kentucky reveals rather vigorous fluctuations. One would surmise that with the new safety legislation in underground mining, existing strip miners would expand operations, thereby improving output per man-day. From 1970, however, a significant decline in productivity values occurred. A partial explanation is that many new strip miners with little experience entered the industry, thereby filling the supply gap left by the closing of underground operations. It is therefore not surprising to find average surface productivity in the industry actually declining. Many of these new strip miners came from other industries, such as construction, in search of new opportunities since construction of roads, dams, and other public facilities was declining nationwide. Capital equipment used in the construction industry is highly substitutable for equipment used in strip mining.

Eastern Kentucky strip mines also may have experienced a decrease in labor productivity in response to 1971 modification in land-reclamation requirements (KRS350) and the enforcement thereof. Enforcement of the legislation is aimed at better restoration and reclamation of land after mining. If the operators are to meet the intent of the new requirements, they have to allocate more labor time to the tasks of preventing landslides, sedimentation, and acid pollution. This results in lower labor productivity. But the productivity picture in surface mining is not at all clear, and there has been considerable controversy over whether in fact strip-mining laws are effectively enforced.[37] And when the values shown in Table 45 are compared with those of 1968, it is evident that strip-mining productivity in 1973 in eastern Kentucky was still more than one-fourth greater than in 1968.

Strip-mining productivity in Ohio and Pennsylvania in 1973 was also higher than in 1968, but by not nearly as much as in eastern Kentucky. In West Virginia, on the other hand, strip-mining productivity in 1973 was actually lower than in 1968.

Whether the regional productivity differences in strip mining are due solely to differences in land-reclamation requirements and their enforcement is uncertain. To be sure, differences in geological and topographical forma-

Figure 4. Productivity in Coal Mining for Four Appalachian Coal Basin States, 1968-1973

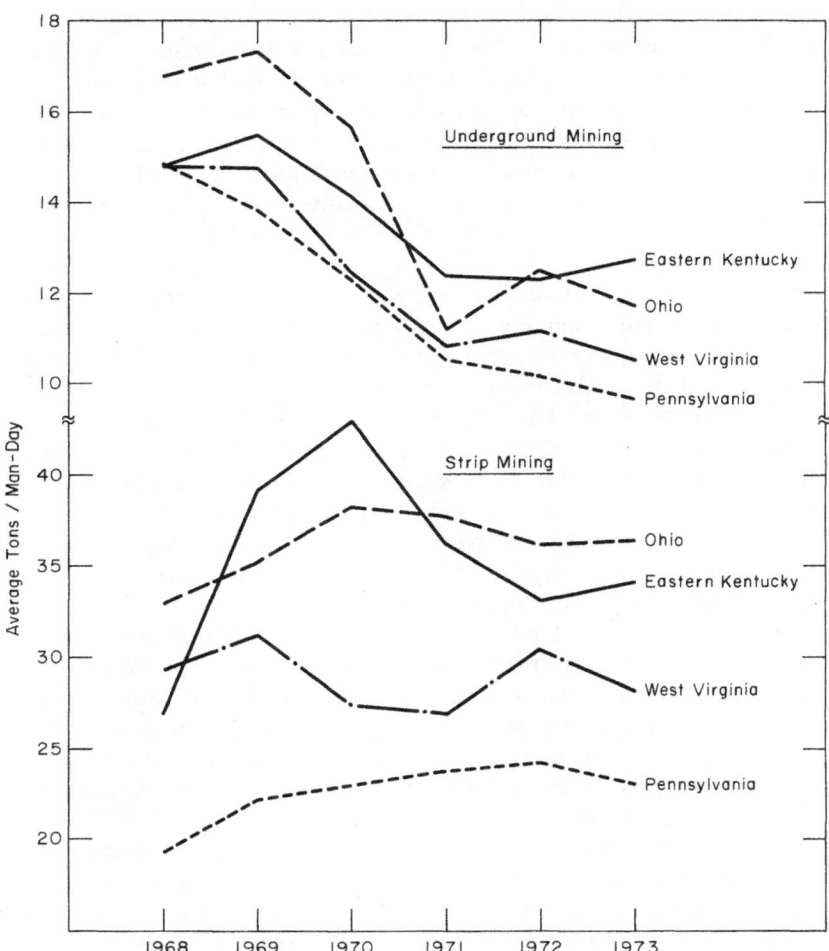

Source: Table 44.

tions of strippable land can have a major impact upon mining productivity. Thus one would expect Ohio strip miners, confronted by gentler terrain than eastern Kentucky miners, to be more productive than the latter, but the values shown in Table 45 do not reflect this. One would also expect West Virginia's surface miners to show productivity patterns similar to those in eastern Kentucky. But once again, statistics in Table 45 show the opposite. It is therefore appropriate to search for an explanation in differing land-reclamation requirements and enforcements among the regions as a major institutional factor responsible for varying productivity values. In addition, consideration should be given to the ease with which strippable reserves can be recovered, that is, to the amount of the overburden that must be removed to get at the coal seam and to the thickness of the seam itself. There is evidence that the most attractive strippable reserves have already been recovered in West Virginia, but less so in eastern Kentucky.[38] As a consequence future gains in strip-mining productivity in eastern Kentucky, given the existing state of mining technology, may not occur.

The absolute productivity values shown in Table 44 indicate that by virtue of its fairly level terrain and very large strip mines, Ohio has the highest productivity values of the four subregions considered. For two years between 1968 and 1973, eastern Kentucky strip miners were actually more productive than their Ohio competitors, but by 1973 their output per manday had once again fallen below that of their northern neighbors.

Pennsylvania and West Virginia strip miners show productivity levels considerably below those of eastern Kentucky and Ohio. When compared with eastern Kentucky, Pennsylvania surface miners were one-third and West Virginia miners one-fifth less productive in 1973. Without careful and more exhaustive study it is only possible to explain these differences with reference to varying land-reclamation requirements in these states and the likelihood that more and better strippable land is available in eastern Kentucky than in West Virginia and Pennsylvania.

A ranking of eastern Kentucky counties in accordance with productivity in strip, auger, and underground mining is shown in Tables 46-48. The data presented in these tables do not reveal a general pattern whereby it can be postulated that mines located in some counties, regardless of whether they are strip, auger, or underground mines, are clearly more productive than mines located in other counties. However, a few counties, among them Lawrence, Martin, Pike, Harlan, and Breathitt, appear to be in the upper ranges of productivity in all three types of mining, while other counties, in particular Owsley, Jackson, Letcher, Whitley, Morgan, Boyd, and Rockcastle, seem to fall into the lower ranges of productivity. Other counties reveal relatively high productivity levels in one type of mining, but only

Table 46. Eastern Kentucky Coal-Producing Counties in Descending Order of Strip-Mining Productivity, 1973

		Average Tons per Man-Day	Annual Output (Thousand Tons)	Number of Mines	Average Size (Annual Tons)
1.	Lawrence	87.87	222	5	44,400
2.	Martin	55.07	3,176	15	211,733
3.	Wayne	53.88	47	2	23,500
4.	Harlan	52.65	727	27	26,926
5.	Pike	45.00	2,821	63	44,778
6.	Breathitt	42.87	4,585	36	127,361
7.	Knott	41.97	491	17	28,882
8.	Floyd	37.26	1,219	23	53,000
9.	Leslie	36.60	609	16	38,062
10.	Elliott	34.98	175	3	58,333
11.	Magoffin	32.62	675	11	61,364
12.	Pulaski	31.17	314	1	314,000
13.	Johnson	30.00	448	19	23,579
14.	Laurel	29.31	230	13	17,692
15.	Morgan	28.75	127	2	63,500
16.	Boyd	28.56	152	4	38,000
17.	Bell	28.45	2,641	16	165,062
18.	Clay	27.82	175	6	29,167
19.	Perry	27.79	2,198	25	87,920
20.	Owsley	26.19	151	5	30,200
21.	Knox	24.80	647	28	23,107
22.	Whitley	21.68	618	30	20,600
23.	McCreary	20.85	138	3	46,000
24.	Carter	20.83	71	4	17,750
25.	Rockcastle	20.74	6	1	6,000
26.	Letcher	19.00	1,004	23	43,652
27.	Jackson	18.70	2	1	2,000

Source: Bureau of Mines, "Coal—Bituminous and Lignite in 1973," *Mineral Industry Surveys,* 1975.

moderate or even low productivity in all others. Leslie, Floyd, and Bell counties in particular illustrate a diversity in productivity levels across two or all three types of mining. For the most part, the more productive counties lie in the easternmost areas of the coal-producing area.

An effort was made to examine whether differences in labor productivity, that is, output per man-day, can be explained in terms of mine size. For that purpose, average mine sizes were calculated for all eastern Kentucky coal counties. Tables 46-48 show the results of these calculations for auger, strip, and underground mining. The data are plotted in Figures 5 and 6.

The relationships developed indicate that with a coefficient of correlation of .58 for auger and .57 for underground mining, mine size is fairly well correlated with output per man-day. Strip mining, on the other hand, with a coefficient of correlation of only .139 shows no significant correlation

Table 47. Eastern Kentucky Coal-Producing Counties in Descending Order of Underground-Mining Productivity, 1973

		Average Tons per Man-Day	Annual Output (Thousand Tons)	Number of Mines	Average Size (Annual Tons)
1.	Leslie	18.84	1,480	14	105,714
2.	Knott	14.54	3,078	48	64,125
3.	Harlan	14.35	8,965	67	133,806
4.	Martin	13.67	2,674	12	222,833
5.	McCreary	13.43	887	4	221,750
6.	Pike	13.30	14,169	210	67,471
7.	Bell	11.51	645	21	30,714
8.	Perry	10.53	2,540	28	90,714
9.	Clay	10.01	145	5	29,000
10.	Letcher	9.46	3,132	59	53,085
11.	Johnson	9.42	148	3	49,333
12.	Floyd	9.35	2,393	103	23,233
13.	Pulaski	9.34	66	2	33,000
14.	Whitley	8.63	194	4	48,500
15.	Lee	5.21	20	1	20,000
16.	Knox	4.19	16	3	5,333

Source: Bureau of Mines, "Coal—Bituminous and Lignite in 1973," *Mineral Industry Surveys,* 1975.

between the two variables. Factors other than mine size influence productivity levels in strip mining.

In eastern Kentucky auger mining in 1973, the largest mine had an annual output of only 57,000 tons, the smallest, 2,000 tons. This means that by expanding the size of operations, substantial increases in productivity can be obtained if all else remains the same. Such increases in productivity would result, of course, in cost savings per unit of output. In underground mining, on the other hand, expanding annual output would generate only a moderate increase in productivity with very small production-cost savings. The slopes of the two regression lines plotted in Figure 5 clearly illustrate the differences in obtainable productivities as a function of mine size for auger and underground mining. In short, economies of scale seem to be operative in auger mining, but only marginally so in underground mining.

In strip mining, it is clear that factors other than mine size control the level of productivity in eastern Kentucky (see Fig. 6). Some of these probably are depth of overburden, width of seam, accessibility to mine, length of mining experience, and type of ownership.

One of the principal determinants of per-unit costs of production, and ultimately price, is the level of productivity. Many factors interact to influence productivity, among them capital-labor ratios, relative prices of inputs,

Table 48. Eastern Kentucky Coal-Producing Counties in Descending Order of Auger-Mining Productivity, 1973

		Average Tons per Man-Day	Annual Output (Thousand Tons)	Number of Mines	Average Size (Annual Tons)
1.	Floyd	90.64	792	28	28,286
2.	Breathitt	54.90	1,769	31	57,064
3.	Harlan	54.48	628	32	19,625
4.	Lawrence	53.31	94	3	31,333
5.	Bell	53.02	441	10	44,100
6.	Clay	49.23	208	5	41,600
7.	Knox	48.45	204	17	12,000
8.	Pike	46.40	2,099	83	25,289
9.	Perry	46.39	1,035	29	35,690
10.	Martin	45.96	304	7	43,428
11.	Laurel	45.09	76	9	8,444
12.	Whitley	41.05	219	20	10,950
13.	Leslie	40.49	313	19	16,474
14.	Wayne	40.35	23	2	11,500
15.	Letcher	38.03	638	28	22,786
16.	Johnson	35.00	272	17	16,000
17.	Knott	33.76	403	23	17,522
18.	Owsley	30.10	72	3	24,000
19.	Magoffin	27.54	114	7	16,286
20.	Boyd	26.23	17	2	8,500
21.	McCreary	22.29	8	1	8,000
22.	Elliott	15.85	2	1	2,000
23.	Morgan	15.41	6	1	6,000
24.	Rockcastle	14.17	3	1	3,000

Source: Bureau of Mines, "Coal—Bituminous and Lignite in 1973," *Mineral Industry Surveys,* 1975.

technological know-how, natural constraints such as size of overburden and width of coal seam, and capacity use. Of these, capacity use is one of the more important.

When compared with mines located in neighboring states, eastern Kentucky mines do not appear to be using their capacity fully. Statistics listed in Table 49 show that during 1972 and 1973, Pennsylvania mines operated at least 20 percent more often than eastern Kentucky mines and West Virginia and Ohio mines at least 9 and 12 percent more. In fact, for the five years examined, eastern Kentucky mines were in operation consistently fewer days than mines in the other states.

A possible explanation for the lower number of days worked in eastern Kentucky is the preponderant number of small mines in the region. Such mines are usually only marginally profitable and are very sensitive to fluctuations in demand and in the spot-market price of coal. Falling prices, typically signaling a slackening in demand or overabundance of supply, could cause

Figure 5. Relationship between Productivity and Mine Size in Underground and Auger Mining in Eastern Kentucky, 1973

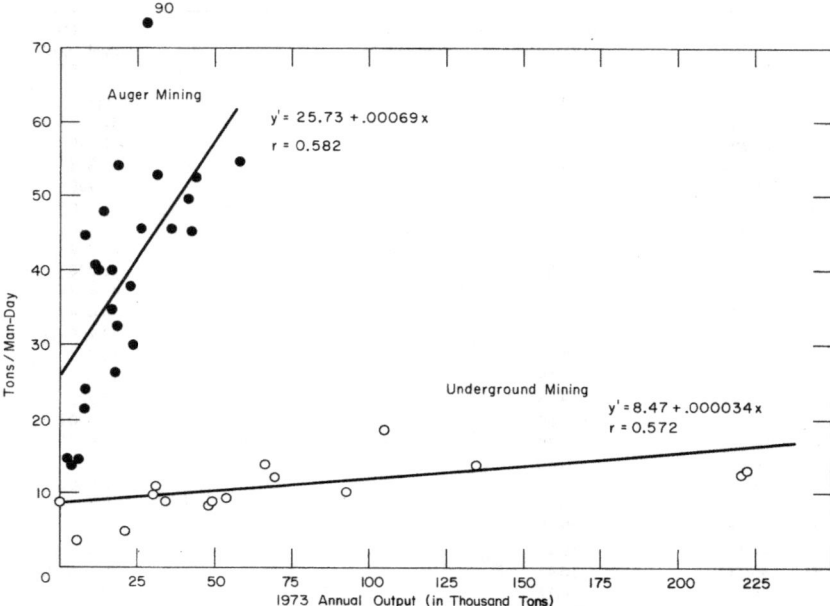

Source: Tables 47 and 48.

them to cease operating. Their fixed investment is very small and in most underground cases was depreciated long ago, so that a temporary closing or even a permanent exit from the industry is not very costly. Larger mines would tend to continue operating in the hope of recouping at least a portion of fixed costs. In the case of small surface mines, most of the equipment may even be rented and fixed costs negligibly small.

Many of the small eastern Kentucky mines are also very responsive to sudden increases in the demand for their coal. The statistics in Table 49 show that the union strike in October and November of 1971 sharply reduced the total number of days mines operated in West Virginia, Pennsylvania, and Ohio. In contrast, the decline in eastern Kentucky was very small. The labor force in the mines is not as heavily unionized in eastern Kentucky as in the neighboring states, so that the national strike had a relatively small impact there. In fact, because of general nonunionization, eastern Kentucky mines showed only a 2 percent decline in the number of days worked in 1971,

Figure 6. Relationship between Productivity and Strip Mine Size in Eastern Kentucky, 1973

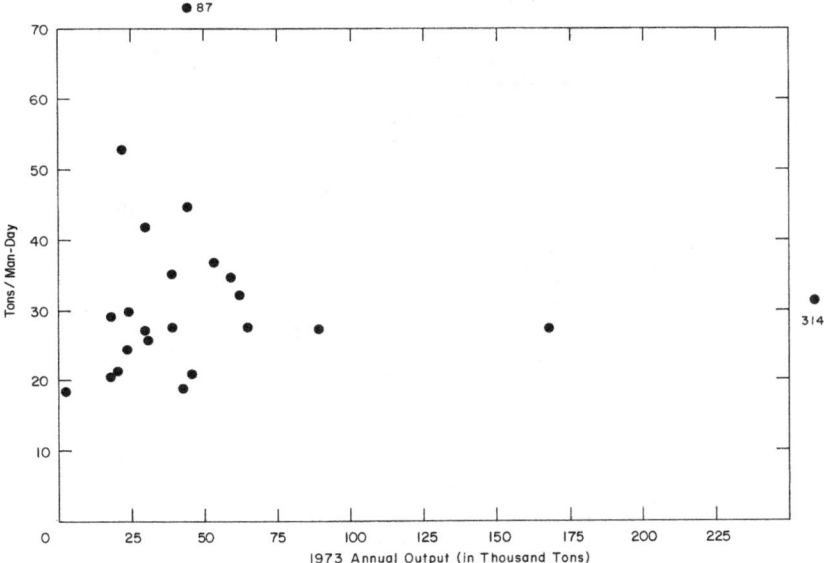

Source: Table 46.

whereas West Virginia, Pennsylvania, and Ohio declined 11, 8, and 3 percent respectively.

Because many eastern Kentucky mines are nonunion, their wage scales are somewhat lower than in union mines. It is often alleged that small mines in eastern Kentucky survive principally because pockets of unemployment continue to exist which exercise an erosive influence on union organization and union-wage levels. Put another way, idle capacity tends to keep mine employment and wages down and unions out. Table 50 shows that unionization in eastern Kentucky was approximately half that of neighboring mining states in 1968. Low wages also keep production costs down, which enables marginal operators to continue to sell coal at competitive prices. Otherwise, so the argument goes, many of these mines would have to cease operations, since they are not as productive as competing mines located elsewhere. While this view had some validity in the 1960s, statistics show that today eastern Kentucky mines are as productive, if not more so, than mines in neighboring states (see Table 44). With other factors constant, it must be argued that if nonunion wage scales are lower in eastern Kentucky than union scales else-

Table 49. Number of Days Worked by Coal Mines
in Four Appalachian Coal Basin States, 1969-1973

	Year	Average Number of Days Worked[a,b]
Eastern Kentucky	1969	191
	70	183
	71	179
	72	192
	73	199
West Virginia	1969	212
	70	219
	71	195
	72	209
	73	218
Pennsylvania	1969	245
	70	248
	71	228
	72	245
	73	240
Ohio	1969	249
	70	235
	71	228
	72	237
	73	222

[a]1969 and 1971 data only published in Bureau of Mines, *Weekly Mineral Industry Survey.*
[b]A labor strike lasted during all of October and part of November of 1971.

Sources: Bureau of Mines, "Coal—Bituminous and Lignite," *Minerals Yearbook,* 1969-1973.

Table 50. Unionization in Coal Mining, 1968

	Production Workers in Mines with Collective Bargaining Agreements Covering a Majority of Such Workers	
	Underground Mines (percent)	Surface Mines (percent)
Eastern Kentucky	45-49	N.A.
Ohio	90-94	40-44
Pennsylvania	90-94	N.A.
West Virginia	90-94	N.A.

Source: Frederick L. Bauer, "Wages in Bituminous Coal Mines," p. 51.

Table 51. Active Mining Days as a Function
of Mine Size in Eastern Kentucky, 1972

Mine Size Category	Percent Under 150 Active Days	Percent Between 150 and 200 Active Days	Percent Over 200 Active Days
1	0	6	94
2	11	15	74
3	26	20	54
4	36	26	38
5	69	14	17
6	94	3	3

Source: Calculated from Kentucky Department of Mines and Minerals, *Annual Report,* 1972.

where, profits of eastern Kentucky mines will tend to be relatively greater. In fact, it has been argued that smaller mining operators often keep capital equipment idle purposely in order to discourage unionization of their employees. The argument implies that even though idle equipment extracts a cost from the mining operator, this cost is smaller than the combined cost of having to pay higher union wages and having to negotiate with an organized group of workers.[39] Whether these arguments are valid is beyond the scope of this survey, but it is possible to observe that in 1972, the smaller mines operated fewer days than the larger mines. Table 51 shows that most of the larger mines operated over 200 days in 1972, whereas most of the smaller mines were active fewer than 150 days. However, more exhaustive study of the reasons for these differences in active days among various size mines is needed before it can be stated with certainty that small mines purposely tend to keep capital equipment idle from time to time in order to discourage union activity.

THE PRICE OF COAL

Economic theory teaches us that with other things equal, there exists an inverse relationship between productivity and price. As productivity increases, price falls. Conversely, as productivity declines, price rises. Of course other things usually do not remain constant in the world around us, but it is correct to argue that increases and decreases in productivity generate downward or upward pressures on price respectively. That is to say, if prices are generally on an upswing, improvements in productivity lead to smaller price rises than would have occurred if productivity had remained constant.

The link between changes in price and productivity is production cost, so that, for example, higher productivity means lower production costs and therefore lower prices or alternatively smaller price increases. Unfortunately, no data are available which would make it possible to separate empirically the effects on price of changing levels of productivity from other factors such as changes in demand and capacity.

Table 52 shows the behavior of prices of underground and strip-mined coal in four subregions of the Appalachian Coal Basin over the past six years for which statistics are available. The general upward trend is unmistakable. Only during 1971 and 1972 was there a moderation in the upswing, a direct consequence of the temporary price controls imposed by the federal administration. Between 1972 and 1973, coal prices resumed their upward movement at an accelerated rate and in 1974 and 1975 prices rose to levels undreamed of only a few years ago. The average U.S. price of coal approximately doubled in 1974, but prices of coal from the Appalachian Basin more than doubled as the demand for this high-quality coal expanded vigorously while supply lagged behind.

The unexpected phenomenon of rapidly rising coal prices in 1974 was a direct consequence not only of sharp increases in the world price of oil but also of the uncertainties created by the foreign oil embargo of 1973. Utilities and other domestic users of oil, confronted with political interruptions in its flow, began to search for substitute energy sources. Coal, one such substitute, became a prime target in the scramble for more secure energy resource supplies. In addition, government agencies encouraged, and in some cases even ordered, the switch from oil- to coal-generated electric power. In 1974 the threat of a labor strike in the coal industry was quite real even in the early part of the year and the ensuing stockpiling of coal put further demand pressures on an industry already operating at capacity.

All these basically short-term phenomena found expression in coal-price increases that were truly remarkable in magnitude. In the short run, capacity to produce coal could not expand rapidly enough to accommodate the burgeoning demand, and buyers eagerly outbid one another in the marketplace.

A manifestation of things to come can be gleaned from the 1973 data in Table 52. In all four subregions 1973 prices increased by at least 12 percent regardless of whether the coal was deep mined or strip mined. Ohio strip-mined coal increased by a remarkable 29 percent in that year. Compared with 1968 levels, nearly all 1973 prices were twice as high.

Eastern Kentucky coal prices in 1973 were higher than 1968 levels for both types of mining; they had risen over that time more rapidly than prices in the other three subregions. In part at least, this is a reflection of the traditionally lower prices of eastern Kentucky coal, Ohio coal exempted. It is clear that eastern Kentucky and Ohio coal operators were demanding and

Table 52. The Price of Coal, 1968-1973

	Year	Underground			Strip		
		Price per Ton	Percentage Change	Price Ratio per Ton 1973/68	Price per Ton	Percentage Change	Price Ratio per Ton 1973/68
Eastern Kentucky	1968	$ 4.63		1.00	$3.35		1.00
	69	5.63	+21.6		3.60	+ 7.5	
	70	7.65	+35.9		5.35	+48.6	
	71	8.89	+16.2		6.19	+15.7	
	72	9.46	+ 6.4		6.23	+ 0.6	
	73	10.63	+12.4	2.30	7.05	+13.2	2.10
Ohio	1968	4.46		1.00	3.72		1.00
	69	4.65	+ 4.3		3.79	+ 1.8	
	70	5.43	+16.8		4.41	+16.4	
	71	6.75	+24.3		4.75	+ 7.7	
	72	7.41	+ 9.8		5.29	+11.4	
	73	8.50	+14.7	1.91	6.82	+28.9	1.83
Pennsylvania	1968	5.97		1.00	3.84		1.00
	69	6.53	+ 9.4		4.24	+10.4	
	70	8.12	+24.4		5.38	+26.9	
	71	9.88	+21.7		6.41	+19.1	
	72	10.39	+ 5.2		6.86	+ 7.0	
	73	12.02	+15.7	2.01	7.68	+11.9	2.00
West Virginia	1968	5.46		1.00	4.31		1.00
	69	5.90	+ 8.1		4.77	+10.7	
	70	8.07	+36.8		7.06	+48.0	
	71	10.07	+24.8		7.49	+ 6.1	
	72	10.90	+ 8.2		7.54	+ 0.6	
	73	12.24	+12.3	2.24	8.58	+13.8	1.99

Sources: Bureau of Mines, "Coal—Bituminous and Lignite," *Mineral Industry Surveys*, 1968-1973.

receiving higher than average prices for their coal, thereby beginning to close the price gap between their coal and that of Pennsylvania and West Virginia.

Figure 7 shows the changes in underground and surface-mined coal prices for the past six years. The trend clearly shows the steep rise in underground coal prices from 1969 on, with only a slight respite in 1971. In 1974 prices climbed to record levels. One explanation for the rise in prices is the effort of coal operators to pass on to the users of coal higher per-unit costs which developed from the installation of the newly required health and safety equipment. Another explanation, in particular for 1974, is the surging demand for coal brought on by the embargo on oil and the quadrupling in its price. The price increases in the four regions were universal and their magnitudes fairly similar.

The underground price increases during 1968-1973 have been much greater than those in strip mining, further exacerbating the competitive cost disadvantage of underground operators. If in subsequent years national strip-

Figure 7. Price of Coal in Selected Appalachian Basin States, 1968-1973

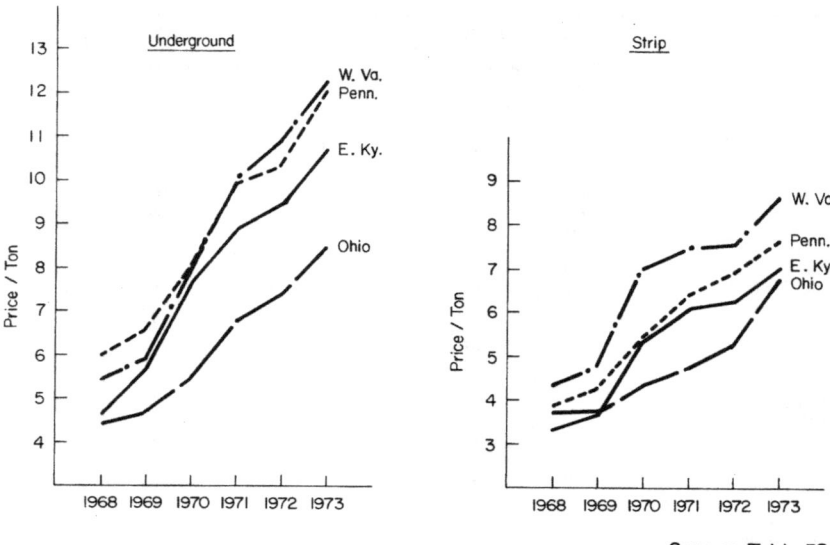

Source: Table 52.

mining legislation is enacted by Congress, surface operators would undoubtedly face additional reclamation costs which in turn would reduce their competitive cost advantage.

To what extent the recent price increases of coal are a reflection of demand, capacity constraints, legislatively imposed cost additions, market concentration, and other factors is difficult to determine. But all these variables affect price, and they do so in varying degrees at various times. The most important factor in 1974 was a spiraling demand set into motion by international political events. Since these subsided, demand moderated and prices have fallen.

No statistics are available which show how much coal is sold in the open or spot market and how much is sold on contract. It is therefore not possible to identify price trends in these two markets separately. But it is true that in the past two years, spot-market prices escalated to levels several times greater than those of the contract market, resulting in a change in coal market shares. Table 53 shows that as the 1972 price in the spot market rose, mining firms producing for the commercial market expanded output. Their share of the market increased significantly while the share of captive mines in total output declined. This was true for all four subregions listed in Table 53. In Ohio, the share of captive mines declined to 2 percent of total output from 9 percent in the preceding year.

Table 53. Percentage of Coal Sold Commercially and by Captive Mines in Four Appalachian Coal Basin States, 1969-1973

	Commercially	Captive Mines
Eastern Kentucky		
1969	88	12
70	89	12
71	89	11
72	95	5
73	89	11
West Virginia		
1969	86	14
70	87	13
71	89	11
72	94	6
73	89	11
Pennsylvania		
1969	63	37
70	63	37
71	65	35
72	75	25
73	69	31
Ohio		
1969	88	12
70	88	12
71	91	9
72	98	2
73	89	11

Source: Table 33.

Statistics dealing with the profitability of individual coal-mining firms operating in eastern Kentucky are unavailable. Most firms consider such information proprietary and do not publish it. Data for large coal firms operating in a multitude of states are more readily available, but they do not reveal profitability breakdowns by geographic origin or type of mining. In addition, many of the large coal firms are wholly owned subsidiaries of enterprises active in diverse noncoal industries and publish consolidated profit-and-loss statements only. It is not possible to extract from these statements data pertaining exclusively to coal operations.

It is common knowledge that the dramatic increase in coal prices in 1974 has spurred an equally dramatic rise in the net incomes of coal companies. Table 54 shows that at the upper extreme, Pittston Coal Company and Westmoreland Coal Company reported 1974 net income at least six times larger than 1973 net income. Only North American Coal and American Electric

Power showed a modest 11 percent and 2 percent increase in 1974 net income. The remaining companies listed in Table 54 show a variety of net income increases over 1973, all representing a new dramatic departure from the experience of previous years.

Selected data for the first half of 1975 show continued strong performance of coal companies. The increases in net incomes of the companies for which data are available vary from a rise of 9 percent to 600 percent. Thus, even though the worldwide recession of 1974-1975 spawned a decline in the demand for coal and in its price, net incomes do not appear to have been affected greatly so far. Prior to the onset of what is commonly referred to as the energy crisis, the coal industry was not particularly distinguished in profitability. But in 1974 the surge in the demand for coal propelled prices upward to levels far in excess of per-unit production costs. Windfall profits were the logical consequence of these events. But this is basically a short-run phenomenon that is not likely to endure as the demand for coal moderates and supply adjustments have time to take place.

THE PROSPECTS FOR EASTERN KENTUCKY COAL

Extravagant and overconfident statements notwithstanding, the long-run future of the eastern Kentucky coal industry is by no means clear.[40] On the one hand, the unpredictability of future oil price levels introduces into forecasts of prospective demand levels for eastern Kentucky coal a considerable amount of uncertainty. On the other hand is the fact that eastern Kentucky coal is of better quality than most, that is, it ranks high in fixed carbon content and low in moisture, volatile matter, and sulfur. It is also located within easy access of its markets, and a large portion of it can be mined with relatively inexpensive surface methods. Its competitive cost position when compared with other coals seems quite favorable.

Nearly two-thirds of eastern Kentucky's coal is used by electric utilities, mostly those located in the South Atlantic, East South Central, and East North Central regions of the United States. These are high-population areas that in the past have demonstrated substantial rates of economic growth. But a great many states in this area are also in a favorable geographic position to substitute oil for coal if relative prices of the two fuels make such substitution economically attractive. They are states that have developed over the past eight to ten years at least some capacity to use oil in addition to or in place of coal.

In 1972 states located in the lower portion of the East Coast purchased over half of the coal shipped from eastern Kentucky to electric utilities. Map 6 shows the geographic location of these states and their percentage share in total residual oil imports. While the New England states historically have relied completely on residual fuel oil as an energy source, they were forced to

Map 6. Geographic Market Areas for Imported Residual Fuel Oil

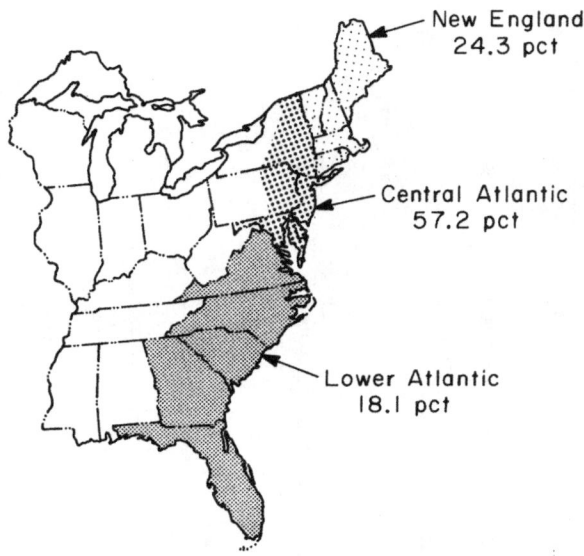

Source: Bureau of Mines, "Comparative Transportation Costs of Supplying Low-Sulfur Fuels to Midwestern and Eastern Domestic Markets," *Information Circular* 8614, 1973.

begin importing coal from abroad as a substitute fuel during the oil embargo of 1973. This situation developed because they were unable to obtain domestic coal on short notice, but were able to buy it from Poland. Given sufficient lead time, however, these states could become coal markets for Appalachian coal if foreign oil becomes unavailable once again.

The demand for the other approximately one-third of eastern Kentucky's annual coal output depends upon the vigor of the steel industry at home and abroad. High-quality coal of the type mined in eastern Kentucky is sought not only by domestic steelmakers but by foreign steel companies as well. [41] If technology holds constant, the fortunes of the steel industry, and therefore of that portion of the eastern Kentucky sector that supplies it with coking coal, will remain closely tied to the general level of economic activity. Worldwide recessions of the type that have beset the industrialized nations in the mid-1970s do not stimulate growth in the steel or the coking coal industry. But essentially these are short-term phenomena that do not appear to engender long-term consequences.

In the near term, the outlook for the eastern Kentucky coal industry seems quite favorable. Since emission control of sulfur oxides and particulate

Table 54. Net Incomes of the Major Coal-Producing Companies in the U.S., 1973-1975

Coal Company	Rank	1973 Net Income (Millions)	1974 Net Income (Millions)	Percent Increase	1975 1st half Profits	Percent Change Over 1st half 1974
Peabody Coal Co. (Kennecott Copper)	1	$159.4	$ 210.9	32	N.A.	
Consolidation Coal Co. (Continental Oil Co.)	2	242.7	327.6	35	$ 60.7	598
Island Creek Coal Co. (Occidental Petroleum)	3	71.9	280.7	291	85.5	94
Amax	4	105.1	148.4	41	N.A.	
Pittston Coal Co.	5	15.3	107.4	600	109.8	181
United States Steel	6	325.8	634.9	95	N.A.	
Bethlehem Mines Corp. (Bethlehem Steel Corp.)	7	206.6	342.0	66	N.A.	
North American Coal	8	4.5	4.9	11	3.7	30
Peter Kiewitt	9	N.A.	N.A.		N.A.	
Old Ben Coal Co. (Standard Oil of Ohio)	10	89.4	147.5	65	N.A.	
Eastern Associated (Eastern Gas & Fuel)	11	17.3	42.0	143	33.4	9
Westmoreland Coal Co.	12	4.7	36.1	669	37.0	166

Pittsburgh & Midway (Gulf Oil)	13	800.0	1,065.0	33	N.A.
Utah International	14	55.4[a]	96.9[b]	75	N.A.
General Dynamics	15	41.1	52.9	29	N.A.
American Electric Power	16	182.6	186.9	2	
Rochester & Pittsburgh	17	N.A.	N.A.		N.A.
Ziegler Coal Co. (Houston Natural Gas)	18	31.1[c]	61.1[d]	96	
C & K Coal Co.	19	N.A.			
Industrial Generating Co. (Washington)	20	N.A.			
Washington Irrigation & Development Co.	21	N.A.			
Kemmerer Coal Co.	22	N.A.			
Falcon Coal Co.	23	3.9	14.4	264	3.9	70

[a]Year ending October 31, 1973   [c]Year ending January 31, 1974
[b]Year ending October 31, 1974   [d]Year ending January 31, 1975

Note: Ranked in accordance with annual production listed in *Keystone Coal Industry Manual*, 1974.

Sources: Calculated from statistics prepared by Research Department, UMWA, and based on *Moody's Industry Manual*, 1974; *Coal Week* (August 11, 1975).

matter is difficult and costly to accomplish where coal is burned, the trend toward using clean low-sulfur coal has accelerated. The eastern Kentucky coal industry with its large low-sulfur deposits has, of course, profited greatly from this trend. It is likely to continue doing so at least until technology to remove stack gases from burned coal is perfected sufficiently so as to find wide acceptance. Some cleaning equipment has already been installed, by Louisville Gas and Electric Company, for example, while others are still under development.[42] It is widely believed that within the next ten years, sulfur-cleansing equipment, rather than very tall smokestacks used for dispersing pollutants, will become accepted for use by electric utilities. If this occurs, high-sulfur coal from the Eastern Interior Coal Basin could become a competitive substitute for the higher priced eastern Kentucky coal. However, high-sulfur coal that must be purged of its sulfur content generates additional costs which arise from the cleansing process. Those who use eastern Kentucky coals would, of course, be able to avoid these costs. It is not possible to predict at this point which of the two types of coal—higher-priced low sulfur or lower-priced high sulfur coal which requires cleaning—will become economically more desirable. In the long run, one would expect to see the emergence of an equilibrium price that reflects not only the different use characteristics of coal but also the additional costs associated with using it in an environmentally responsible manner.

3. The Western Kentucky Coal Industry

The western Kentucky coal industry, which differs greatly in character and structure from its eastern Kentucky counterpart, is a major sector of the Kentucky economy. In 1975 western Kentucky mines produced 53 million tons of coal and provided employment for more than 9,000 men.[1] Nearly one-fourth of all persons employed in Kentucky mining worked in western Kentucky. In several counties, the importance of coal mining is greater than state statistics indicate. For example, in Muhlenberg County, the coal industry provided 28 percent of total 1974 employment.

The coal-producing counties of western Kentucky are part of the large Eastern Interior Coal Basin which also covers a portion of Indiana and most of Illinois (see Map 7). The coals found in the basin are relatively homogeneous in that they exhibit similar calorific values and produce similar amounts of dioxides and particulate matter. Table 55 shows the results of analyses made by the United States Bureau of Mines on 1974 tipple and delivered samples taken at various points in the basin. The differences in state regions are small and it is therefore correct to speak in terms of a homogeneous resource when describing coal mined in the basin, whatever its geographic origin.

Of particular interest in recent years has been the sulfur content of coals, mostly as a consequence of concern over air purity. To a lesser extent the same is true of the particulate matter or ash contained in coal when it is emitted into the atmosphere during combustion. All the coal mined in the basin contains substantial amounts of sulfur and therefore contributes significantly to air pollution. This is also the case, although to a lesser extent, of particulate matter. The relatively clean-burning coals of eastern Kentucky and West Virginia contain amounts of ash similar to those found in coal of the Eastern Interior Coal Basin.

The BTU content of coals mined throughout the basin is also very similar, and variations in BTU values per pound from region to region are small. Table 55 shows that of the 1973 samples taken in the basin, western Kentucky coal contained the highest BTU values by at least 3.2 percent over the next highest state sample—Indiana. Illinois samples contained the lowest BTU values.

Map 7. Eastern Interior Coal Basin

Source: Richard C. Neavel, *Petrographic and Chemical Composition of Indiana Coal,* Geological Survey Bulletin No. 22 (Bloomington: Department of Conservation, 1961).

Whether these differences are due entirely to a conscious effort to mine only the highest-value coal or to the type of coal actually found in western Kentucky is difficult to establish, but previous Bureau of Mines analyses revealed a similar difference between western Kentucky and Indiana-Illinois coals.[2]

In terms of output, western Kentucky ranks fifth in coal production in the nation. In 1975, 53 million tons of coal were mined in the area, approximately 8 percent of national output or 38 percent of the Kentucky total. Table 56 shows the size of and the fluctuations in western Kentucky output over time in relation to fluctuations in U.S. output.

With the exception of 1971, output in western Kentucky has remained remarkably constant over the past nine years. The sharp upsurge in coal mined in 1970 was the result of several special influences. First, wildcat-labor strikes in West Virginia increased the demand for western Kentucky coal.

Table 55. Analysis of Tipple and Delivered Samples in Western Kentucky, Indiana, and Illinois, 1973
(percent per ton)

Origin of Coal Sampled	Approximate Ash Content	Approximate Sulfur Content	Approximate BTU/dry
Western Kentucky	6.93	3.21	13,446
Indiana	9.45	2.10	13,030
Illinois	9.96	2.40	12,878

Note: All figures are weighted average estimates and no attempt was made to segregate run-of-the-mine from washed coal samples.

Source: Bureau of Mines, *Analysis of Tipple and Delivered Samples of Coal.*

Table 56. Coal Production in the U.S. and Western Kentucky, 1967-1975
(thousands of tons)

	United States	Percent Change	Western Kentucky	Percent Change
1967	552,000		46,390	
1968	545,000	−1.3	46,515	+0.03
1969	560,000	+2.8	47,466	+2.0
1970	603,000	+7.7	52,803	+11.2
1971	552,000	−8.4	47,825	−10.4
1972	595,000	+7.8	52,330	+10.5
1973	591,000	−0.7	53,679	+2.6
1974	602,000	+1.9	51,841	−3.4
1975	640,000	+6.3	53,034	+2.3

Sources: Bureau of Mines, *Minerals Yearbook,* 1967-1975; "Weekly Coal Report 3056," *Mineral Industry Surveys,* 1976; "Coal—Bituminous and Lignite," *Mineral Industry Surveys,* 1973-1975.

Table 57. Share of Western Kentucky, Indiana, and Illinois in National Output, 1967-1975
(percentage of total)

	Western Kentucky	Illinois	Indiana
1967	8.4	11.8	3.4
1968	8.5	11.5	3.4
1969	8.5	11.6	3.6
1970	8.8	10.8	3.7
1971	8.7	11.0	3.9
1972	8.8	11.0	4.4
1973	9.1	10.4	4.3
1974	8.6	9.6	3.9
1975	8.3	9.3	3.8

Sources: Bureau of Mines, *Minerals Yearbook,* 1967-1975; "Weekly Coal Report 3056," *Mineral Industry Surveys,* 1976; "Coal—Bituminous and Lignite," *Mineral Industry Surveys,* 1973-1975.

Second, a general upsurge in the electric utility demand for coal affected all regions not plagued by strikes. Third, inventories were being built up in 1970, mostly in anticipation of the general 1971 strike. Fourth, surface-mined coal improved its cost-price position relative to deep-mined coal because the new federal safety regulations caused the cost of deep-mined coal to escalate sharply. Fifth, the new, more restrictive air-pollution laws were not as yet being implemented in 1970, and therefore, high-sulfur, high-ash western Kentucky coal continued to be viewed as an acceptable substitute—albeit a temporary one—for cleaner and more expensive coal mined elsewhere. The output of western Kentucky coal in 1975 was approximately 14 percent greater than the 1967 output, consistent with the general pattern of only modest growth (approximately 1.7 percent per year). This performance is slightly below that recorded by the industry as a whole, for which output in 1975 was 16 percent greater than in 1967.

In terms of the percentage share of western Kentucky output in national coal production during the past eight years, Table 57 shows that the region's share has not changed greatly, fluctuating from 8.3 to 9.1 percent, but did fall in 1975 to its lowest level in nine years.

Indiana's output also has been relatively stable, while the Illinois share of

Table 58. Total Value of Coal Mined in the U.S., Western Kentucky, Illinois, and Indiana, 1967-1974
(millions of dollars)

	United States	Percent Change	Western Kentucky	Percent Change	Illinois	Percent Change	Indiana	Percent Change
1967	$2,555		$159		$253		$ 73	
1968	2,546	0	159	0	250	− 1.0	72	− 2.3
1969	2,796	+ 9.8	167	+ 5.0	280	+11.7	83	+15.7
1970	3,775	+35.1	222	+33.3	320	+14.6	103	+23.7
1971	3,887	+ 2.9	231	+ 4.1	331	+ 3.5	111	+ 7.6
1972	4,561	+17.3	274	+18.6	402	+21.4	145	+30.6
1973	5,049	+10.7	318	+16.1	413	+ 2.7	153	+ 5.5
1974	9,015*	+78.5	462	+45.3	582	+40.9	198	+29.4

*Preliminary estimate.

Sources: Bureau of Mines, *Minerals Yearbook*, 1967-1971; "Weekly Coal Report 3029," *Mineral Industry Surveys*, 1975; "Coal—Bituminous and Lignite," *Mineral Industry Surveys*, 1973-1974.

output over the period has declined somewhat. It is fair to state that none of the three regions has shown significant fluctuations in output when compared with national production and that western Kentucky and Indiana have maintained their relative positions during the past nine years.

The total value of coal produced during the past eight years in western Kentucky, Indiana, Illinois, and the nation is shown in Table 58. The inordinate surge in demand for coal during 1970 (and again in 1972) abated markedly in 1971, as did the pressure on price. But despite the nationwide decline in tons of coal mined during 1971, the total value of this coal increased. The entire 1971 increase in the value of coal mined in the basin was due to the continued rise in the price of coal exclusive of transport costs (FOB mine). Between 1972 and 1974, statistics show that it was less the change in tonnage mined and more the rapid rise in price that propelled upward the total value of coal sold. This is particularly so for 1974. Even though United States output rose by only 2 percent, and western Kentucky production actually declined, escalating coal prices pushed up total 1974 values 78 percent nationally and 45 percent for western Kentucky. These price surges were precipitated by political events that transpired in the world petroleum markets in 1973 and may be viewed as temporary economic rents obtained by the coal

Table 59. Comparative Values of U.S., Western Kentucky, Indiana, and Illinois Coal, 1967-1974
(dollars per ton, FOB mine)

	United States	Percent Change	Western Kentucky	Percent Change	Indiana	Percent Change	Illinois	Percent Change
1967	$ 4.62		$3.42		$3.91		$ 3.88	
1968	4.67	+ 1.0	3.42	0	3.88	− 0.8	4.01	+ 3.3
1969	4.99	+ 6.8	3.52	+ 2.9	4.13	+ 6.4	4.30	+ 7.2
1970	6.26	+25.5	4.22	+19.8	4.60	+11.4	4.92	+14.4
1971	7.04	+12.5	4.83	+14.4	5.18	+12.6	5.46	+11.0
1972	7.66	+ 8.8	5.23	+ 8.3	5.58	+ 7.7	6.14	+12.4
1973	8.53	+11.4	5.93	+13.4	6.06	+ 8.6	6.71	+ 9.3
1974	15.00*	+75.8	8.92	+45.4	8.36	+37.9	10.00	+49.0

*Preliminary estimate.

Sources: Bureau of Mines, *Minerals Yearbook*, 1967-1971; "Weekly Coal Report 3003," *Mineral Industry Surveys*, 1975; "Coal—Bituminous and Lignite," *Mineral Industry Surveys*, 1973-1974.

industry. The magnitude of these rents, however, astounded many observers.

Table 59 shows the changes in the price of coal between 1967 and 1974. For the entire eight-year period, per-ton price increases for the basin fell short of the national increases. During this time, the national price more than tripled, while for western Kentucky, Indiana, and Illinois, prices only more than doubled. The main explanation for this difference is the strong increase in the demand for low-sulfur coal with upward pressure on their prices. This type of coal is not characteristic of the coal mined in the basin, hence the disparity in price movements between the basin and the nation as a whole.

In 1973 primary employment in the coal-mining industry of the Eastern Interior Coal Basin was 20,656 workers. Of these, the largest number, 10,500, were employed in Illinois, 7,476 in western Kentucky, and 2,680 in Indiana. Table 60 shows employment in the basin and its breakdown by region and type of mining between 1967 and 1973. The figures reveal that there has been a considerable change in total employment in western Kentucky during the period, an increase of 47 percent, while in Illinois and Indiana the increase was somewhat smaller, 31 and 45 percent respectively. In both western Kentucky and Indiana there occurred an increase in surface-mining employment of 40 and 74 percent respectively. In underground

Table 60. Average Number of Men Working Daily in Coal Mines in Western Kentucky, Illinois, and Indiana, 1967-1973

		Underground	Surface	Total
Western Kentucky	1967	3,100	1,973	5,073
	1968	3,313	1,864	5,177
	1969	2,036	3,000	5,036
	1970	2,720	2,519	5,239
	1971	3,885	2,696	6,581
	1972	4,514	2,896	7,410
	1973	4,716	2,760	7,476
Illinois	1967	4,881	3,107	7,988
	1968	5,158	3,247	8,405
	1969	5,133	3,244	8,377
	1970	6,057	3,220	9,277
	1971	6,481	3,117	9,598
	1972	7,068	3,256	10,324
	1973	7,229	3,271	10,500
Indiana	1967	446	1,405	1,851
	1968	440	1,725	2,165
	1969	463	1,534	1,997
	1970	487	1,786	2,273
	1971	478	1,885	2,363
	1972	363	2,450	2,813
	1973	240	2,440	2,680

Sources: Bureau of Mines, *Minerals Yearbook,* 1967-1973; "Coal—Bituminous and Lignite," *Mineral Industry Surveys,* 1973-1974.

mining, employment either declined, as in Indiana, or as in western Kentucky, rose even more rapidly than in surface mining. In Indiana, this change is a reflection of the shift in coal mining from underground to surface operations, but not in western Kentucky. In Illinois, on the other hand, the increase in total employment came primarily from growth in underground rather than surface mining. Employment in the former rose 48 percent in the seven years while it changed practically not at all in surface mining.

In sum, growth in underground-mining employment is occurring only in western Kentucky and Illinois. In surface mining, employment growth is occurring in western Kentucky and Indiana, but not in Illinois. Employment in Indiana underground mining is declining. It should be noted that at least a portion of the increase in underground-mining employment in western Kentucky and Illinois is directly attributable to the recently strengthened safety precautions now mandatory in underground mines. It is also probable that employment in Indiana underground mines would have fallen even more steeply than it has, if federal safety legislation had not been implemented.

The statistics considered thus far concern primary employment only, that

Table 61. Total Primary and Secondary Employment
in the Eastern Interior Coal Basin in 1973
(multiplier effect of 1.9)

Western Kentucky	14,204
Illinois	19,950
Indiana	5,092

Source: Calculated from Table 60.

is, persons employed directly in the industry. They do not include employment which arises outside of the industry as a direct consequence of the former's existence.

Primary and secondary employment combined represent the total number of jobs created by the coal industry and gives a more accurate picture of the total employment impact of the industry. A recent study shows that such employment in transport, service, electric power, and other industries can escalate direct employment by an estimated 90 percent.[3] If a base multiplier of 1.9 is assumed appropriate for the entire basin, total employment in each region would, of course, nearly double. Table 61 shows this employment for each of the three regions in the basin during 1973.

Western Kentucky is well endowed with recoverable coal reserves. Estimates of the size of reserves vary from optimistic Bureau of Mines figures to somewhat more conservative Tennessee Valley Authority (TVA) figures.[4] Ted D. Haley of the University of Kentucky Institute for Mining and Minerals Research compiled the set of independent reserve estimates for western Kentucky counties; these are shown in Table 62. Nearly 85 percent of the estimated reserves are located in five counties—Henderson, Hopkins, Muhlenberg, Union, and Webster. Table 63 separates these reserves into those which are recoverable through strip mining and those recoverable by other methods such as underground mining. The table shows that if strippable reserves are considered separately, the first three counties dominate, while Webster County's coal reserves are recoverable principally with underground-mining methods.

Together, the counties of Henderson, Hopkins, Muhlenberg, and Ohio account for nearly three-fourths of western Kentucky's strippable coal reserves. If we disregard Henderson County, where at present no strip mines are being operated, the life of the reserves, at 1971 production rates, extends from an estimated nine years in Muhlenberg to thirty-two years in Ohio County. Clearly, Hopkins and Ohio counties offer the greatest promise for continued large-surface operations in the long run. The level of stripping coal

Table 62. Total Recoverable Reserves in Short Tons and Percentage of the Western Kentucky Coal District Recoverable Reserves per County

County	Total Recoverable Reserves (Strip and Underground)	Percent of Western Kentucky Reserves
Breckinridge	NA	NA
Butler	38,377,000	0.51
Caldwell	NA	NA
Christian	21,704,000	0.29
Crittenden	7,000,000	0.09
Daviess	143,250,000	1.92
Edmonson	8,610,000	0.02
Grayson	1,000,000	0.01
Hancock	13,927,000	0.19
Henderson	1,328,158,000	17.81
Hopkins	1,403,288,000	18.82
McLean	437,340,000	5.86
Muhlenberg	1,171,626,000	15.71
Ohio	458,569,000	6.16
Union	1,487,965,000	19.95
Webster	936,845,000	12.56
Total	7,457,659,000	99.90

Sources: Ted D. Haley, "A Kentucky Coal Utilization Research Program," *Annual Report,* Project 1: *Geology and Economics* (Lexington, Ky.: Institute for Mining and Minerals Research, 1974); Bureau of Mines, "Strippable Reserves of Bituminous Coal and Lignite in the United States," *Information Circular* 8531, 1971; TVA, *Coal Reserves of Western Kentucky;* Kentucky Department of Mines and Minerals, *Annual Report,* 1972.

in Muhlenberg County, on the other hand, presages an early end to operations, while the largest single untapped source of strippable coal lies in Henderson County.

The life of reserves subject to underground removal is much greater than that for strippable reserves. Enough coal remains in the ground to assure hundreds of years of mining at 1971 production rates.

Comparable data for recoverable strip reserves and estimated production

Table 63. Western Kentucky Coal District Coal Reserves and Life Beyond 1971 at 1971 Rates of Production

County	Approximate Reserves (Thousand Tons)	Strippable Reserves				Underground Reserves			
		Percent of Reserves	1971 Production	Life Beyond 1971 (Years)	Recoverable Reserves (Thousand Tons)	Percent of Reserves	1971 Production	Life Beyond 1971 (Years)
Butler	38,377	4.48	161,014	238			64,497	
Caldwell			0				0	
Christian	6,743	0.79	165,951	41	14,961	0.23	0	
Crittenden	7,000	0.82	0				0	
Daviess	65,146	7.61	1,093,772	60	78,104	1.18	0	
Edmonson	8,610	1.01	390,094	22			0	
Grayson	1,000	0.12	0				0	
Hancock	13,927	1.63	73,000	191			0	
Henderson	165,000	19.28	0		1,163,158	17.62	69,187	16,812
Hopkins	132,805	15.52	5,224,321	25	1,270,483	19.24	6,255,825	203
McLean	42,432	4.96	474,353	89	394,908	5.98	0	
Muhlenberg	183,227	21.41	19,287,534	9	988,399	14.97	4,058,266	244
Ohio	141,883	16.58	4,441,522	32	316,686	4.80	1,431,623	221
Union	25,000	2.92	0		1,462,965	22.16	2,947,344	496
Webster	24,565	2.87	636,538	39	912,280	13.82	1,050,199	867
TOTAL	855,715	100.00	31,948,099	27	6,601,944	100.00	15,876,941	416

Sources: Same as Table 62.

Table 64. Estimated Remaining Coal Reserves of the Eastern Interior Coal Basin, January 1, 1974
(by sulfur range, in million short tons)*

	≤1.0	1.1–3.0	>3.0	Unknown	Total
Illinois	1,095.1	7,341.4	42,968.9	14,256.2	65,664.4
Indiana	548.8	3,305.8	5,262.4	1,504.1	10,621.1
Western Kentucky	0.2	564.4	9,283.9	2,815.4	12,664.4

*Data may not add to totals shown due to rounding.
Source: Bureau of Mines, "Demonstrated Coal Reserve Base of the United States."

lives for Indiana and Illinois are unavailable, but an estimate of total remaining coal reserves in the United States by type, sulfur content, and location shows that by far the largest bituminous coal reserves of any state are located in Illinois.[5] Indiana and western Kentucky reserves are approximately equal in size; but even combined, they comprise less than half of the reserves found in Illinois, as can be seen in Table 64. The table also shows that considerably more than half of the reserves contain at least 3 percent sulfur and that only an insignificant amount of the coal is sufficiently clean to meet new air-purity standards without some type of purification. In short, coal from the Eastern Interior Basin cannot meet clean-air standards if these are enforced.

THE DEMAND FOR WESTERN KENTUCKY COAL

The demand for coal, a derived demand, is the result of the uses to which coal is put. The demand for these uses, in turn, is derived from the demand for the various end products and services in whose production these uses are inputs. Total demand for coal typically is segregated into intermediate use categories such as the ones shown in Table 65. For example, in 1973, 64.8 percent of total United States production was used by electric utilities and 16.6 percent by coke and gas plants.

The table also shows the extreme dependence of the Eastern Interior Coal Basin on electric utilities as its principal customer. For western Kentucky, this dependence is even greater than for the basin as a whole. No less than 94 percent (50 percent more than the national average) of western Kentucky's 1973 coal output was purchased by public utilities for use in the generation of electric power. The percentage of Indiana- and Illinois-produced coal consumed by utilities is smaller than western Kentucky's, but also significantly

Table 65. Origins of Demand for Coal Mined in the U.S.
and Eastern Interior Coal Basin, 1973
(percentage)

	United States	Western Kentucky	Illinois	Indiana
Electric utilities	64.8	94.0	80.2	81.7
Coke and gas plants	16.6	0.0	7.2	0.0
Retail dealers	1.4	0.4	1.1	0.4
All others	11.0	5.5	12.5	17.9
Overseas exports	6.0	0.0	0.0	0.0
Miscellaneous	0.2	0.1	0.0	0.0
	(100.0)	(100.0)	(100.0)	(100.0)

Source: Bureau of Mines, "Bituminous Coal and Lignite Distribution, Calendar Year 1974," *Mineral Industry Surveys,* 1975.

above the national figure. These statistics suggest that the fortune of the western Kentucky coal industry is closely related to that of the electric power industry, in particular that portion located in western Kentucky, Tennessee, and the other southeastern states.

The origins of the demand for electric power are industrial, residential, and commercial in nature (see Table 66), and, as Table 67 shows, electric utilities are consuming a steadily increasing share of total coal output. Little change in the continued growth of demand for electric power is likely to occur in the future. The nation's demand is expected to continue to expand in the 1970s and 1980s, propelled largely by the expanding needs of an increasingly affluent society.[6] Electricity has the virtues of being a clean, versatile, and relatively inexpensive source of power when compared with available substitutes. With incipient natural gas shortages, this expansion will probably continue well into the future.

Forecasting the demand for coal more than a year or two in advance is at best a hazardous undertaking. But estimates abound nonetheless. An estimate of the future demand for electricity, which is the source of more than 90 percent of the demand for western Kentucky coal, is shown in Table 9. A more modest set of estimates is presented in Table 68, and one of the more conservative of these is the Electric Power Research Institute (EPRI) projection. The reason for the modest nature of the EPRI estimate (and that of the Federal Energy Administration [FEA] estimate which is based on EPRI figures) is the fact that it incorporates sectorial price elasticities of demand

Table 66. Categories of Electric Power Use in the U.S., 1965-1990

Category	Percent of Total Use		
	1965	1970	1990
Industrial	41	40	41
Residential	24	25	24
Commercial	18	18	20
Miscellaneous, including losses	17	17	15
	(100)	(100)	(100)

Source: Federal Power Commission, *1970 National Power Survey*, pt. 1.

for electric power while the other estimates generally treat the demand for electricity as price-insensitive.

A graphic representation of the electric utility demand for coal, based on a set of independent estimates, is shown in Figure 8. By 1980 these estimates predict that more than 500 million tons of coal will be consumed by electric utilities, an average annual increase of approximately 5 percent. Whether this growth rate will actually be maintained, exceeded, or remain unrealized is, of course, uncertain. These forecasts are only tentative guidelines. But it is correct to anticipate that the demand for coal over the long run will continue to grow, particularly if devices to purge coal of its polluting impurities are fully perfected and installed, so that high-sulfur coal can be used without the risk of degrading the environment.

The various sources of the demand for coal from the Eastern Interior Basin and the changing relative importance of these sources between 1969 and 1973 is shown in Table 69. The most significant upward change occurred in the demand from electric utilities. This is clearly reflected in their steadily growing share in the coal market. At the same time, the share of consumers of bituminous coal other than electric utilities, coke and gas plants, retail dealers, railroads, coal mines, and mine employees (as defined by the Bureau of Mines) has declined sharply in western Kentucky and Indiana and less dramatically in Illinois. Such users are usually largely industrial plants. Despite this decline, these users continue to absorb about one-fifth of Indiana's total output and about one-eighth of Illinois's, but only one-twentieth of western Kentucky's output. There is little doubt, however, that the demand for basin coal from the electric utilities will continue to grow more rapidly than the demand from any other user for some time into the future.

Table 67. Share of Electric Utilities in Total Coal Consumed in the U.S. and Eastern Interior Coal Basin, 1967-1973
(percentage)

	United States	Western Kentucky	Illinois	Indiana
1973	64.8	94.0	80.2	81.8
1972	63.8	93.2	79.1	77.7
1971	61.8	91.3	78.3	75.8
1970	56.7	89.6	75.0	70.5
1969	55.7	86.8	73.2	68.0
1968	54.5	86.6	71.5	67.9
1967	53.6	86.3	69.5	64.6

Sources: Bureau of Mines, *Minerals Yearbook*, 1967-1972; "Bituminous Coal and Lignite Distribution, Calendar Year 1974," *Mineral Industry Surveys*, 1975.

Table 68. Comparison of Forecasts on U.S. Electricity Consumption
(trillions of kwh)

		EPRI	FEA[1]	NPC[2]	NERC[3]	ITC[4]	CTM[5]	EPP[6]
	High	2.85	2.67	3.78			3.05	
1980	Medium	2.71	2.56	3.70	2.74	2.72	2.55	
	Low	2.56	2.44	3.56			2.05	
	High	3.83	3.72	4.71				3.78
1985	Medium	3.50	3.46	4.59	3.83	3.55	2.78	2.34
	Low	3.16	3.20	4.33				2.31

Note: Actual Electricity Consumption

1960	0.758
1965	1.055
1970	1.532
1975	1.901

[1] Federal Energy Administration, *Project Independence Report*, November 1974.
[2] National Petroleum Council, *U.S. Energy Outlook: A Summary Report of the National Petroleum Council*, December 1972.
[3] NERC Forecast released July 1975. Extended from 1984 to 1985 by Curtis E. Harvey.
[4] Intertechnology Corporation, *The U.S. Energy Problem*, November 1971. These numbers were compiled from thirty-five separate forecasts.
[5] Chapman, Tyrrell, and Mount, "Electricity Demand Growth and the Energy Crisis," *Science* 178 (November 1972): 703-8.
[6] *A Time to Choose*. Energy Policy Project of the Ford Foundation, 1974.

Source: Electric Power Research Institute, *A Preliminary Forecast of Energy Consumption through 1985* (Palo Alto, Calif.: 1976), p. 20.

Table 69. Origin of Demand for Coal Mined in the Eastern Interior Coal Basin, 1969-1973
(percentage)

User	Indiana					Illinois					Western Kentucky				
	1969	1970	1971	1972	1973	1969	1970	1971	1972	1973	1969	1970	1971	1972	1973
Electric utilities	68.0	70.5	75.8	77.7	81.7	73.2	75.0	78.3	79.1	80.2	86.8	89.6	91.3	93.2	94.0
Coke and gas plants	0.6	0.9	0.0	0.3	0.0	5.0	6.8	6.9	6.4	7.1	0.4	0.1	0.2	0.0	0.0
All others[a]	30.1	26.0	21.6	20.7	17.9	19.1	16.1	13.3	12.6	12.5	10.1	8.8	7.4	6.3	5.5
Overseas exports	0.0	0.0	0.0	0.0	0.0	0.0	0.0	0.0	0.0	0.0	0.0	0.0	0.0	0.0	0.0
Retail dealers	1.1	1.7	1.3	1.1	0.4	2.2	1.8	1.5	1.5	1.0	2.4	1.5	0.8	0.3	0.3
Miscellaneous	0.2	0.9	1.3	0.2	0.0	0.5	0.3	0.0	0.4	0.0	0.3	0.0	0.3	0.3	0.2
Total	100.0	100.0	100.0	100.0	100.0	100.0	100.0	100.0	100.0	100.0	100.0	100.0	100.0	100.0	100.0

[a]Includes mostly shipments to industrial plants and consumers of bituminous coal other than electric utilities, coke and gas plants, retail dealers, railroads, coal mines, and mine employees.

Sources: Bureau of Mines, *Minerals Yearbook*, 1969-1972; "Bituminous Coal and Lignite Distribution, Calendar Year 1974," *Mineral Industry Surveys*, 1975.

Figure 8. Demand for Coal by Electric Utilities

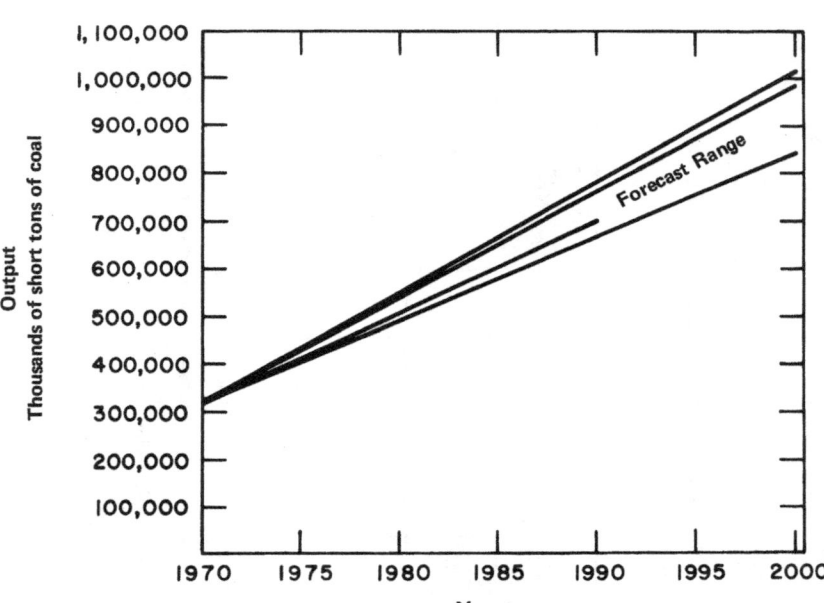

Source: Curtis E. Harvey, *The Western Kentucky Coal Industry in 1974* (Lexington: Institute for Mining and Minerals Research, 1975).

During recent years, the need for energy resources to produce electric power has grown more rapidly than any other need for such resources, at least for most of the early 1970s. Figure 9 compares growth in energy consumption with growth in population. Except for the depression years, the diagram shows how rapidly demand for energy has grown in this century and how the gap between it and growth in population continues to widen. Westinghouse Corporation estimated that in order to satisfy at least some of the incipient future demand for energy, the United States between 1970 and 1990 will need new electricity-generating capacity of 1,000 gigawatts (GW, or 10^9 watts). This is twice the present installed capacity of roughly 480 GW.[7] Where this additional capacity will come from is uncertain. Possibilities include nuclear, solar, tidal, geothermal, and wind sources. But it is clear that none of these sources alone will be sufficiently developed and operational in the next twenty years to be able to satisfy all the additional energy needed for the production of electric power. Collectively, however, they can be expected to make a significant contribution.

Figure 9. The Rate of Growth in U.S. Energy Consumption and Population, 1800-2000

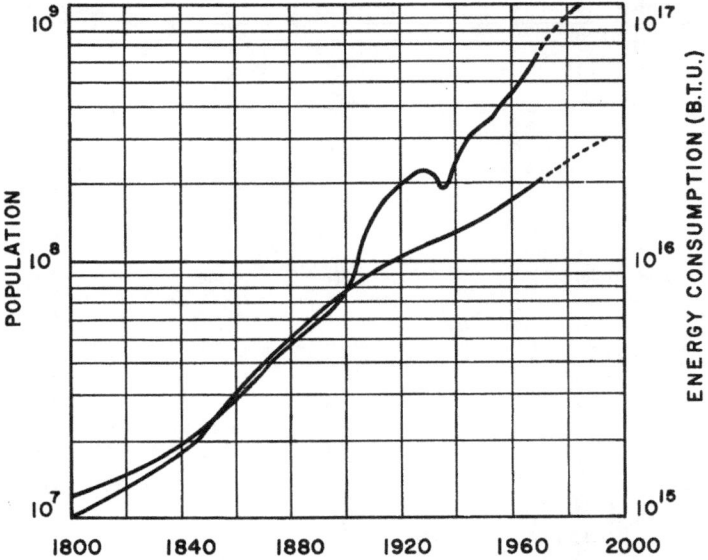

Source: *Scientific American* 224 (September 1971).

During 1973 electric public utilities located in Indiana, Illinois, and Kentucky consumed 79,952,000 short tons of coal, an amount equal to approximately 57 percent of basin output. Not all this coal originated in the basin, although the major share did. Similarly, not all the coal mined in the basin and destined for electric public utilities is shipped to points within the confines of the area. Nevertheless, since at least three-fourths of the demand for coal mined in the basin originated with such utilities, it is appropriate to focus the market analysis on this type of demand.

Nearly 80 percent of the coal mined in Indiana is shipped to destinations located within the state. The remainder of the output is distributed over nine other states, with none receiving more than 6 percent of the state's total output (see Table 70). It is apparent, therefore, that Indiana mines produce coal mostly for internal use and that most of this coal (79 percent in 1973) is shipped to electric utilities located in the state. The remaining coal used by power plants in Indiana comes principally from western Kentucky and Illinois. On balance, Indiana is an importer of coal.

Less of the coal mined in Illinois is consumed in the state (not quite one-

Table 70. Market Shares in Output of the Eastern Interior Coal Basin, 1973

Consuming State	Percent of Total Illinois Output	Percent of Total Indiana Output
Ohio	0.0	1.7
Illinois	47.2	1.6
Indiana	9.7	78.5
Kentucky	4.7	1.8
Michigan	1.7	1.3
Alabama and Mississippi	2.1	0.0
Tennessee	1.4	1.9
Georgia and Florida	1.2	6.2
Missouri	15.5	0.0
Wisconsin	8.5	4.8
Minnesota	2.7	1.0
Iowa	6.3	0.1
Total	100.0	100.0

Source: Bureau of Mines, "Bituminous Coal and Lignite Distribution, Calendar Year 1974," *Mineral Industry Surveys*, 1975.

Table 71. Major Markets for Indiana and Illinois Coal, 1973
(percentage of output)

Indiana to		Illinois to	
Indiana[a]	78.5	Illinois[b]	47.2
Wisconsin	4.8	Missouri[b]	15.5
		Wisconsin	8.5
		Indiana	9.7
		Iowa[b]	6.3
Total	83.3		87.2

[a] Markets in which Indiana sells considerably more than the other states in the region
[b] Markets in which Illinois sells considerably more than the other states in the region

Source: Table 70.

Table 72. Market Shares of Illinois and Eastern Interior Coal Basin

State	1973 Consumption (Thousands of tons)	Percent of Total Consumption Originating in	
		Illinois	Eastern Interior Basin
Missouri	17,385	54.9	56.3
Wisconsin	12,634	41.5	71.2
Iowa	6,889	56.3	65.2
Three-State Total	36,908	50.6	63.1
Illinois	40,628	71.6	76.9
Indiana	45,061	13.3	70.3
Five-State Total	122,597	43.8	70.4

Source: Bureau of Mines, "Bituminous Coal and Lignite Distribution, Calendar Year 1974," *Mineral Industry Surveys,* 1975.

half, as the statistics in Table 70 show), but Illinois electric utilities purchase from Illinois producers nearly 75 percent of the coal they consume. An additional 5 percent originates in western Kentucky. The rest is purchased from Wyoming, Montana, and Washington. Once again it appears that the electric utilities are strongly dependent on indigenous coal, presumably because of transport-cost advantages and proximity to the source of the coal.

Table 71 shows the major markets for Illinois and Indiana coal in 1973. The data illustrate once again that in both Indiana and Illinois the largest amount of coal consumed by far is coal mined internally or in contiguous areas. Distance obviously plays an important role in determining market distribution, although an increasing amount of Rocky Mountain coal is being shipped into the state, mostly because the costs of surface mining west of the Mississippi are very low.

The share of the national market served by Indiana and Illinois coal producers is shown in Table 72. For Illinois, statistics reveal that it supplies the major portion of coal consumed in Missouri, Wisconsin, and Iowa. If Indiana and western Kentucky are added as suppliers, then nearly two-thirds of the coal consumed in these states originates in the basin.

As one might suspect, if Indiana and Illinois are added as consumers as well, then, with their sizable internal markets, nearly three-fourths of the coal consumed in these five states originates in the basin, a substantial amount indeed. In short, the Eastern Interior Coal Basin is not only its own principal supplier of coal but also the principal supplier of coal used in several

Map 8. Regional Market Distribution of Western Kentucky Coal, 1973

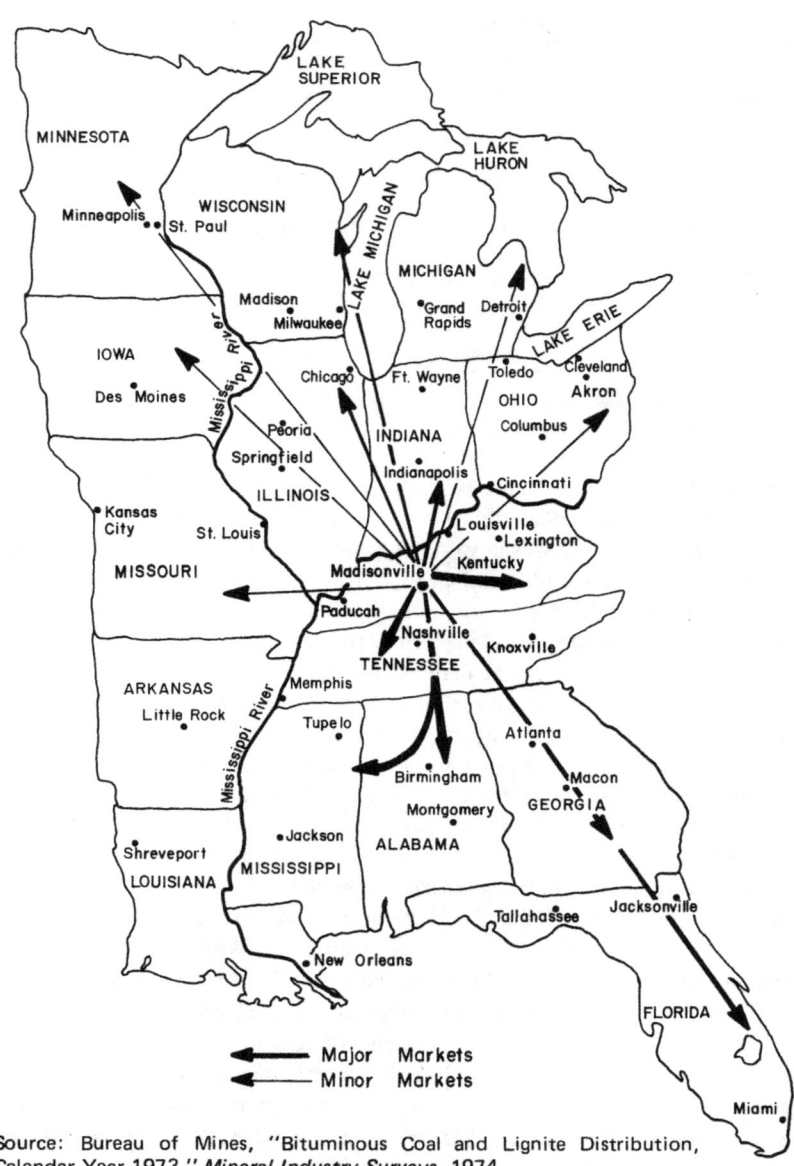

Source: Bureau of Mines, "Bituminous Coal and Lignite Distribution, Calendar Year 1973," *Mineral Industry Surveys,* 1974.

contiguous states. Transportation-cost differences play a major role in delineating the geographic confines of the market.

Because the principal buyers of western Kentucky coal are the electric utilities, we are most interested in the factors that influence the size and regional origin of this demand.

In 1973, 94 percent of the coal mined in western Kentucky was shipped to electric utilities (see Table 65). Eighty-four percent of western Kentucky coal was consumed in Kentucky, Georgia, Florida, Alabama, Mississippi, Indiana, and Tennessee; that is, nearly three-fourths of the market for western Kentucky coal is located either in Kentucky or to the south of it (see Map 8).

The ranking of each state as a market for western Kentucky coal is shown in Table 73. As expected, Kentucky itself is its own best customer, with large amounts of coal being shipped to the TVA power plants located in the state. This market is from one-half to three times the size of the next four market groupings, all of which individually consume between 10 and 20 percent of western Kentucky's annual output. This represents a fairly generous market dispersion, which implies that a decline in the coal market in any one of these states would not severely handicap the western Kentucky industry as a whole.

Table 73 also shows the approximate dependence of each state on western Kentucky coal, but only insofar as this dependence can be inferred from the percentage share of the market which western Kentucky coal holds in each state. For instance, 66.2 percent of the coal used in Kentucky in 1973 came from sources located in western Kentucky, the highest percentage of any of the preceding five years and approximately the same as in 1970. The implication is that more coal from outside the state was used in Kentucky in 1971 and 1972 than in any of the three other years.

Western Kentucky supplies a very significant, though variable, share of the coal used in Tennessee—slightly more than one-half. In Alabama, Mississippi, Georgia, and Florida, there has occurred over the years a small increase in the percentage share of western Kentucky coal in total coal used; in Indiana and Wisconsin there has been a slight decline. The implication of these statistics is that as western Kentucky's coal production has increased between 1967 and 1973 (output in 1973 was 15 percent greater than in 1967), so has its share in its major markets. Only in Indiana and Wisconsin has western Kentucky's share declined somewhat.

Statistics for the past five years show that western Kentucky's coal markets continue to be fairly well dispersed geographically. Alabama, Mississippi, Georgia, Florida, Tennessee, and Kentucky purchase from one-fourth to two-thirds of the coal they consume from western Kentucky coal mines.

Table 73. Markets for Western Kentucky Coal, 1969-1973
(share of market in output and share of coal in total market consumption)

	Percent of Western Kentucky Output Consumed					Percent of Consumption Originating in Western Kentucky				
	1973	1972	1971	1970	1969	1973	1972	1971	1970	1969
Kentucky	29.5	30.3	29.0	29.0	26.7	66.2	57.9	53.7	65.4	61.8
Tennessee	20.0	17.2	12.3	14.2	12.1	50.6	42.1	30.8	41.3	34.2
Alabama and Mississippi	13.0	12.7	14.0	11.5	13.0	26.4	22.2	24.0	22.6	24.0
Georgia and Florida	11.6	14.1	14.1	11.4	8.9	38.6	41.6	40.9	46.0	35.5
Indiana	10.4	12.2	13.8	12.8	14.1	12.9	13.8	16.9	16.2	16.2
Wisconsin	4.5	5.5	6.2	6.7	7.5	19.9	19.2	19.2	20.8	23.9
Illinois	3.2	3.2	3.4	4.9	8.2	4.4	4.1	4.1	6.1	8.6
Michigan	2.7	1.8	2.0	2.3	2.1	4.8	2.8	2.9	3.4	2.7
Ohio	2.7	2.7	3.6	5.0	4.3	2.3	2.1	2.7	4.0	3.3
Iowa	1.0	0.2	0.2	0.6	0.9	8.5	1.8	1.5	5.4	7.9
Minnesota	0.6	0.4	0.8	1.3	1.3	3.8	2.6	4.8	7.9	7.7
Missouri	0.4	0.0	0.0	0.0	0.0	1.4	0.1	0.0	0.0	0.0
Canada	0.2	0.1	0.1	0.1	0.1	0.6	0.4	0.0	0.0	0.0

Sources: Bureau of Mines, "Bituminous Coal and Lignite Distribution, Calendar Years 1970-1974," *Mineral Industry Surveys*, 1971-1975.

To these states, this coal represents an important source of energy. Whether western Kentucky coal producers will maintain or enlarge their positions in these markets in the future or succumb to competition from substitute fuels and/or substitute coals is, of course, highly uncertain. Much depends upon the speed with which equipment to purge coal of sulfur and other impurities is introduced by buyers in the various markets. Recent pronouncements by TVA, Louisville Gas & Electric, and others hold out the promise of large-scale and successful modification of present power-generating equipment for removing ash particles from smokestacks. TVA has announced plans to spend $270 million on pollution-control equipment designed to reduce sulfur dioxide emissions at all eleven of its steam plants by the end of 1977, and Louisville Gas & Electric already has such equipment operating.[8] The wide-scale implementation of sulfur and ash removal could greatly enhance the well-being of the western Kentucky region and of the industry as a whole.

Finally, Table 74 ranks the various coal markets in terms of the percentage of western Kentucky coal output consumed during 1969-1973. Noteworthy is the fact that Tennessee and Kentucky markets have steadily expanded their share in consuming western Kentucky coal, whereas the Alabama, Mississippi, Georgia, Florida, and Wisconsin rankings have remained relatively unchanged. Only Indiana's share has declined in importance. The first five areas represent the traditional markets for western Kentucky coal, and a substantial reduction in coal shipped to any one of these, for whatever reason, could have a detrimental impact on the industry. But there is little reason to suspect that such a reduction in demand will occur in the short run. The cost advantages that derive principally from the barge method of shipping coal for now enable western Kentucky operators to compete effectively in these markets. Tightened EPA restrictions without the installation of sulfur-removal equipment, however, could become a source of a future decline in the demand for western Kentucky coal.

THE SUPPLY OF WESTERN KENTUCKY COAL

This section describes the evidence gathered concerning the output of coal from different size mines and firms concerning relative productivities and access to transport ways. Data are presented and used to illuminate cost conditions found in the coal industry during the period studied. Given these conditions, conclusions are reached concerning the nature of the supply characteristics for coal.[9]

Coal is not commercially mined with the same intensity everywhere it is found. In fact, only forty-three of the total 120 Illinois, Indiana, and western Kentucky counties in which coal is found actually are involved in mining.

Table 74. Rank-Size Distribution of Markets for Western Kentucky Coal, 1969-1973
(by percentage of output consumed)

	Rank	Cumulative 1973 Percent	Rank	Cumulative 1972 Percent	Rank	Cumulative 1971 Percent	Rank	Cumulative 1970 Percent	Rank	Cumulative 1969 Percent
Kentucky	1	29.5	1	30.3	1	29.0	1	29.0	1	26.5
Tennessee	2	49.5	2	47.5	5	41.3	2	43.2	4	38.6
Alabama and Mississippi	3	62.5	4	60.2	3	55.3	4	54.7	3	51.6
Georgia and Florida	4	74.1	3	74.3	2	69.4	5	66.1	5	60.5
Indiana	5	84.5	5	86.5	4	83.2	3	78.9	2	74.6
Wisconsin	6	89.0	6	92.0	6	89.4	6	85.6	7	82.1
Illinois	7	92.2	7	95.2	8	92.8	8	90.5	6	90.3
Michigan	8	94.9	9	97.0	9	94.8	9	92.8	9	92.4
Ohio	9	97.6	8	99.7	7	98.4	7	97.8	8	96.7
Iowa	10	98.6	11	99.9	11	98.6	11	98.4	11	97.6
Minnesota	11	99.2	10	100.3	10	99.4	10	99.7	10	98.9
Missouri	12	99.6	13	100.3	13	99.4	13	99.7	13	98.9
Canada	13	99.8	12	100.4	12	99.5	12	99.8	12	99.0
Total*		99.8		100.4		99.5		99.8		99.0

*Note: Totals do not sum to 100 due to rounding.

Source: Table 73.

Table 75. Production of Coal, Eastern Interior Coal Basin, 1974[a]

	Strip Mining		Underground Mining		Total	
	Millions of Tons	Percent of Total	Millions of Tons	Percent of Total	Millions of Tons	Percent of Total
Illinois	26.9	33.9	31.2	57.6	58.2	43.5
Indiana	23.6	29.8	0.1	0.2	23.7	17.7
Western Kentucky	28.8	36.3	22.9	42.2	51.8	38.8
Total	79.3	100.0	54.2	100.0	133.7	100.0

[a]Production figures are preliminary estimates subject to later change.
Source: Bureau of Mines, "Weekly Coal Report 3029," *Mineral Industry Surveys*, 1975.

Table 75 presents production figures for 1974 by state and by type of mining method. The differences between output proportion and estimated reserves are substantial. Overall, western Kentucky, believed to contain only about 8 percent of total basin reserves, contributes over 38 percent of total basin output, Illinois 43 and Indiana 18 percent. Table 76 shows that during the past eight years, western Kentucky has steadily increased its share in basin output while that of Illinois continues to decline. Indiana's share appears to have reached its peak in 1972. Nationally, the Eastern Interior Coal Basin continues to supply somewhat more than one-fifth of the output with little change observable over the 1967-1974 period.

Another way of expressing this difference between output and reserves is to calculate the theoretical remaining life of the coal supply, that is, the number of years it would take to mine all known reserves, assuming a constant rate of production equal to the current rate.[10] Values using 1970 production figures for underground mining and 1971 figures for surface operations are shown in Table 77. As is apparent, western Kentucky is being mined much more intensively than the other states, particularly in surface operations. In fact, productivity of surface-mined coal-lands has begun to decline, as has output per surface mine.[11]

Carried to the county level, this type of analysis reveals that mining operations in several counties have theoretical lives considerably shorter than the state average of twenty-eight years (see Table 78). One county in particular, Muhlenberg, is noteworthy, since it has a remaining theoretical strip-mining life of only nine years. This implies that a drastic decline in surface mining could take place in that county within the next decade, if 1971 production rates persist. The underground life, however, is much longer—an estimated 244 years.

Table 76. Distribution of Coal Production among U.S. Districts, Eastern Interior Coal Basin, 1967-1974

	Western Kentucky		Illinois		Indiana		Total	
	Quantity Produced (Thousands of Tons)	Percent of Regional Total	Quantity Produced (Thousands of Tons)	Percent of Regional Total	Quantity Produced (Thousands of Tons)	Percent of Regional Total	Quantity Produced (Thousands of Tons)	Percent of National Output
1967	46,390	35.6	65,133	50.0	18,772	14.4	130,295	23.6
1968	46,515	36.5	62,441	49.0	18,486	14.5	127,442	23.4
1969	47,466	35.9	64,722	48.9	20,086	15.2	132,274	23.6
1970	52,803	37.7	65,119	46.4	22,263	15.9	140,185	23.3
1971	47,819	37.5	58,402	45.8	21,396	16.8	127,617	23.1
1972	52,330	36.4	65,523	45.6	25,949	18.0	143,802	24.2
1973	53,679	38.2	61,572	43.8	25,253	18.0	140,504	23.7
1974	51,841	38.7	58,216	43.5	23,726	17.7	133,783	22.2

Sources: Bureau of Mines, *Minerals Yearbook,* 1970-1973; "Weekly Coal Report 3029," *Mineral Industry Surveys,* 1975; "Coal—Bituminous and Lignite," *Mineral Industry Surveys,* 1973-1974.

Since coal mined in Indiana, Illinois, and western Kentucky is taken from the same coalfield, it is useful first to examine the industry of the entire region before focusing on the unique features that characterize western Kentucky mining.

Although there were 175 mines operating in 1973 to supply more than 140 million tons of output, all mines clearly did not contribute equally. In all three states it was the large mines, those with annual outputs of over 500,000 tons, which supplied most of the coal. In 1973 these large mines produced 130 million tons, 93 percent of the region's total, in spite of the fact that only approximately one in every two mines was classified as large. The twenty largest mines in the basin, their 1973 output and location, and percentage share in total production are shown in Table 79. The greatest number of large mines is located in Illinois, the smallest number in Indiana. Together, the twenty largest mines produced 46 percent of the basin's output.

While this information, at first glance, suggests that the industry has many sellers, a more accurate measure of concentration is the distribution of output over firms rather than over the units of production (mines). Table 80 lists fourteen firms operating mines in the region and shows that these fourteen contribute about 80 percent of the region's output. Peabody Coal, the largest supplier, operates a total of twenty-one mines in the region, all of

Table 77. Years of Theoretical Life of Coal Reserves,
Eastern Interior Coal Basin

	Surface Operations	Underground Operations
Illinois	109	2,072
Indiana	53	7,770
Western Kentucky	28	416

Note: Estimates assume production continuing at 1971 rates.

Sources: Calculations based on data from Bureau of Mines, *Minerals Yearbook,* 1973; Ted D. Haley, "A Kentucky Coal Utilization Research Program."

Table 78. Theoretical Life of Strippable Coal Reserves
for Selected Counties in Western Kentucky

County	Theoretical Life (Years)
Muhlenberg	9
Edmonson	22
Hopkins	25
Ohio	32
Webster	39

Source: Ted D. Haley, "A Kentucky Coal Utilization Research Program."

which are in the 500,000-tons-per-year size category. This company produced one-third of the coal mined in the basin in 1973. Peabody's pervasiveness is hardly an isolated phenomenon, as Table 80 shows.

Also evident in the industry is the existence of many small mining operations. Logically, it would seem that small operations would expand to become large producers over time, but this does not seem to be the case. Rather, large operations appear always to have been large (some of the large mines listed in Table 80 began operations as late as 1968) and small operations, always small. Large mines and small mines are essentially different operations. Both types are represented in fairly substantial numbers in the basin.

The United States Bureau of Mines periodically publishes estimates of labor productivity in the coal industry on a county basis. This section

Table 79. The Largest Mines in the Eastern Interior Coal Basin, 1973

Name	Output (Tons)	State	Percent of Regional Total
River King	6,526,267	Ill.	4.64
Sinclair	5,290,991	W. Ky.	3.76
Captain	4,451,313	Ill.	3.16
River Queen	4,172,223	W. Ky.	2.96
Peabody No. 10	4,147,069	Ill.	2.95
Lynnville	4,064,910	Ind.	2.89
Ayrgem	3,206,242	W. Ky.	2.28
Ken	3,202,350	W. Ky.	2.27
Universal	3,043,781	Ind.	2.16
Leahy	2,942,035	Ill.	2.09
Monterey No. 1	2,694,505	Ill.	1.91
Camp No.1	2,619,513	W. Ky.	1.86
Inland	2,588,482	Ill.	1.84
Homestead	2,448,696	W. Ky.	1.74
Vogue	2,412,445	W. Ky.	1.71
Eros	2,396,288	Ind.	1.70
Old Ben No. 24	2,337,482	Ill.	1.66
Orient No. 3	2,207,429	Ill.	1.57
Old Ben No. 26	2,100,316	Ill.	1.49
Wright	2,096,526	Ind.	1.49
20 mines	64,948,863	—	46.22
175 mines	140,504,000	—	—

Note: This list includes 48.7 percent of Illinois output, 45.9 percent of Indiana output, and 44.1 percent of the Western Kentucky output.

Sources: National Coal Association, *Bituminous Coal Facts,* 1970; *Keystone Coal Industry Manual,* 1974.

Table 80. Coal Output by Parent Firms in the Eastern Interior Coal Basin, 1973

Parent Firm	Coal Operating Company	Output	Percent of Region	Cumulative Percent
Kennecott Copper	Peabody	47,574,779	33.86	33.86
American Metal Climax	Amax	12,558,181	8.93	42.79
Standard Oil, Ohio	Old Ben	10,373,737	7.38	50.17
General Dynamics	United Electric, Freeman	8,184,165	5.82	55.99
Ashland Oil	Southwestern Ill.	5,897,046	4.19	60.18
Occidental Petroleum	Island Creek	5,779,198	4.11	64.29
Continental Oil	Consolidation	5,756,564	4.09	68.38
Houston Natural Gas	Ziegler	3,006,256	2.13	70.51
Gulf Oil	Pittsburgh & Midway	2,952,216	2.10	72.61
Exxon	Monterey	2,694,505	1.91	74.52
Inland Steel	Inland Steel	2,588,482	1.84	76.36
Amarron Coal	Cimarron	1,639,838	1.16	77.52
Mapco	Webster County Coal	1,633,596	1.16	78.68
American Smelting & Refining	Midland	1,105,777	0.72	79.40
14 largest companies		111,654,340		79.40
Region		140,504,000		100.00

Source: *Keystone Coal Industry Manual*, 1974.

includes an analysis of these productivity figures for 1970 and comments on labor-productivity differences between large and small mines and between types of mining.

The Bureau of Mines reports output of coal per man-day for counties where coal is commercially mined. In each case, the productivity figure is reported for surface and for underground mining. In a few cases when only one mine is operative in a particular county, the figures for output and productivity refer to that particular mine. More often, more than one mine operate in a county; for these, aggregate or average figures are reported.

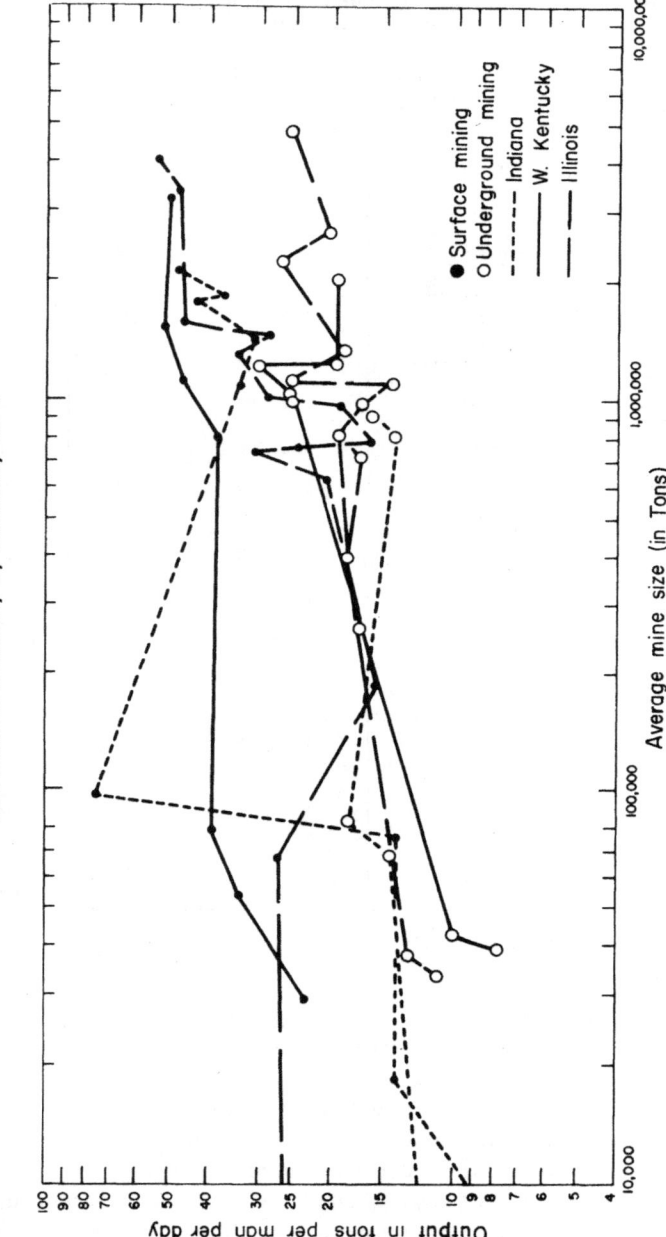

Figure 10. Labor Productivity in Mining for Indiana, Illinois, and Western Kentucky by Counties, 1970

Source: *Minerals Yearbook*, 1972.

To analyze the relationship between size and productivity an effort was made to estimate average mine size by county, in order to compare this with reported average productivity figures. A simple average of size, the arithmetic mean, was not appropriate because productivity is not a simple average. Consequently, a system of weighted averages was employed. While this may introduce some error into the size measure, this approach is superior to other measures and does not unduly influence the validity of conclusions.

Figure 10 illustrates the apparent relationship between size and productivity by type of mining, based on values taken from Table 81 which also indicates whether the figures are derived from actual mine statistics (where no error is likely) or from an estimate of size. Not unexpectedly, what is apparent is that surface operations have a higher labor productivity than underground operations, and regardless of type of mining, larger mines tend to report higher labor productivity than smaller mines.

It would seem that state boundaries in the Eastern Interior Coal Basin are incidental to the economics of coal; however, this appears to be only partially true. The three states in the region exhibit certain unique features which make mining a slightly different industry in each, in terms of size distribution and labor productivity.

Tables 82 and 83 illustrate the size distribution of mining in the three states. Only slight differences are apparent. Illinois is populated with the largest mining operations, with more than three out of every five mines classified as large, that is, over 500,000 tons annually. In contrast, Indiana has only one in three mines classified as large and Kentucky, two out of five. At the other size extreme, Illinois had only 13 percent of its mines in the category of those producing less than 50,000 tons annually, a figure that compares with 39 percent for Indiana and 25 percent for western Kentucky. Thus Illinois has the highest percentage of large mines and the lowest of small mines. The reverse is true for Indiana, with western Kentucky occupying a middle position.

Differences in percentage of total output originating in the largest and smallest size classes are less pronounced within the basin. Illinois and Indiana derive 95 and 92 percent of their output from the largest mines; western Kentucky, approximately 90. At the other extreme, about 0.6 percent of the total of western Kentucky output is produced by its smallest mines, just a little more than Illinois, while the smallest mines in Indiana produce 1.2 percent.

In terms of weighted aggregate averages, western Kentucky coal mines report higher levels of labor productivity than mines located in Illinois and Indiana (see Table 81). This difference in productivity can be attributed largely to the geological features of the western Kentucky coal region.

Data for underground mines reveal that Illinois and western Kentucky

Table 81. Average Size and Productivity of Labor for Mines in Indiana, Illinois, and Western Kentucky, 1970 Production Year

County	Number and Type of Mines	Average Size (Thousand Tons)	Average Productivity (Tons/Man-Day)
		Indiana	
Clay	6 surface	1,082[a]	33.48
Fountain	1 surface	18	14.63
Gibson	1 underground	802	18.93
Greene	4 surface	1,793[a]	37.97
Parke	1 surface	8	8.38
Pike	1 underground	68	14.55
Spencer	1 surface	75	14.32
Sullivan	2 underground	902[a]	16.90
	3 surface	1,736[a]	42.97
Vermillion	1 surface	953	75.11
Vigo	1 underground	82	18.56
Warrick	1 underground	8	12.35
	7 surface	2,180[a]	48.85
State	6 underground	—	17.56
	24 surface	—	40.51
Total	30 mines		36.07[b]
		Illinois	
Christian	1 underground	4,900	25.36
Douglas	1 underground	1,140	25.85
Franklin	4 underground	2,250[a]	27.46
Fulton	6 surface	1,450[a]	29.20
Gallatin	2 underground	1,150[a]	14.10
	1 surface	749	24.74
Jackson	2 surface	67[a]	27.95
Jefferson	3 underground	2,650[a]	20.53
Kankakee	1 surface	976	18.14
Knox	1 surface	1,528	46.17
Macoupin	1 underground	262	17.29
Mercer	1 underground	36	14.20
	1 surface	6	108.16
Montgomery	2 underground	1,372[a]	19.04
Peoria	3 surface	1,018[a]	29.40

County	Mines	Size	Value
Perry	3 surface	3,333[a]	48.21
Pope	2 surface	6[a]	25.14
Randolph	1 underground	827	19.65
	2 surface	1,383[a]	34.40
St. Clair	3 underground	461[a]	19.00
	3 surface	4,000[a]	53.76
Saline	3 underground	724[a]	17.33
	4 surface	789[a]	16.29
Stark	1 surface	622	20.31
Vermillion	2 underground	33[a]	11.88
	1 surface	185	15.47
Williamson	4 underground	950[a]	17.33
	2 surface	746[a]	31.09
State	28 underground	—	21.25
	33 surface	—	33.58
Total	**61 mines**		**26.11**

Western Kentucky

County	Mines	Size	Value
Butler	2 underground	42[a]	10.19
	4 surface	79[a]	39.51
Christian	4 surface	29[a]	23.08
Daviess	1 surface	804	37.42
Henderson	3 underground	39[a]	7.76
	14 underground	1,247[a]	19.69
Hopkins	19 surface	1,156[a]	46.87
McLean	3 surface	53[a]	33.55
Muhlenberg	7 underground	1,071[a]	25.27
	18 surface	3,221[a]	49.81
Ohio	1 underground	1,214	30.72
	13 surface	1,570[a]	51.11
Union	4 underground	1,927[a]	19.55
Webster	1 underground	1,089	25.04
	1 surface	442	n.a.
State (Western Region)	32 underground	—	21.61
	63 surface	—	48.55
Total	**95 mines**		**33.28**

[a]Estimate of size. Actual size not reported. [b]Weighted by size of mine

Sources: Calculations derived from Bureau of Mines, *Minerals Yearbook*, 1972; and other sources.

Table 82. Size Distribution of Mines among the Eastern Interior
Coal Basin States, 1973

	Over 500,000 Tons		Less than 50,000 Tons		State Totals	
	Number of Mines	Output (Thousand Tons)	Number of Mines	Output (Thousand Tons)	Number of Mines	Output (Thousand Tons)
Illinois						
Surface	18	28,749	5	47	32	29,002
Underground	18	31,937	2	51	23	32,570
Total	36	58,686	7	98	55	61,572
Indiana						
Surface	12	22,618	15	298	36	24,465
Underground	1	614	-	-	3	789
Total	13	23,232	15	298	39	25,253
Western Kentucky						
Surface	17	27,748	18	267	55	31,337
Underground	16	20,713	2	35	26	22,342
Total	33	48,462	20	302	81	53,679
Total	82	130,380	42	698	175	140,504

Source: Bureau of Mines, "Coal—Bituminous and Lignite," *Mineral Industry Surveys,* 1974.

Table 83. Percentage Distribution of Mines by Size of Output,
Eastern Interior Coal Basin, 1973

	Over 500,000 Tons		Less than 50,000 Tons	
	Percent of Mines	Percent of Output	Percent of Mines	Percent of Output
Illinois				
Surface	56.2	92.2	15.6	0.2
Underground	78.3	98.0	8.7	0.2
Total	65.4	95.3	12.7	0.2
Indiana				
Surface	33.3	92.4	41.7	1.2
Underground	33.3	77.8	0.0	0.0
Total	33.3	91.9	38.5	1.2
Western Kentucky				
Surface	30.9	88.5	32.7	0.8
Underground	61.5	92.7	7.7	0.2
Total	40.7	90.3	24.7	0.6
Total	46.8	92.8	24.0	0.5

Source: Calculated from Table 82.

mines differ only slightly, while Indiana mines report considerably lower levels of output per man-day for this type of mining (see Table 81). In part this difference appears to result from different sizes of mines operating in the three states. Average productivity is a weighted average giving more weight to large outputs than to small outputs. When this feature is combined with the fact that larger mines are more productive, the statewide average depends to a great extent on the existence of large mines. Indiana has only two underground mines that produce over 500,000 tons of coal per year, and these mines are small relative to the size of mines in the other states. Both mines produce less than a million tons, while both Illinois and Kentucky can boast of several mines with outputs running to several million tons per year. This would tend to bias their state average upward.

Surface mining in Kentucky clearly shows higher productivity than in other states, but a careful examination of the figures in Table 81 reveals that only two counties in Kentucky actually exceed the state average productivity level, measured in tons per man-day. The interesting point is that these two counties, Muhlenberg and Ohio, contribute about 75 percent of the strip output for western Kentucky. This leads one to conclude that it is basically productivity in these two counties which is responsible for the high productivity average of the state. If output from these two counties should fall relative to the other counties, average productivity for the region would decline drastically. The next highest productivity figure comes from Hopkins County. Together these three adjoining counties contribute 92 percent of western Kentucky's surface output.

There exist three main methods of estimating the costs of producing a unit of output. These methods are a statistical cost analysis which uses actual information on production costs and size of output, engineering cost estimates of producing at various levels of output, and a method known as the survival technique which attempts to infer the character of costs from examining the extant size distributions of firms. No method is clearly the best, and selection of one over another is often dictated by the type and amount of data available. For the case at hand, some data were available for each method, so that all three were examined briefly for possible use.

In order to use a meaningful statistical analysis to estimate production costs for coal mining, more information than is now generally available would be required. Some data concerning labor productivity are available, but the link between productivity and cost is not always clear. In general, when firms use identical production techniques, those that enjoy higher labor productivity also have lower costs. However, in cases where underlying production techniques differ, knowledge about productivity is, at best, a very weak indicator of cost conditions. Such is the case with coal.

Surface operations tend to have higher labor productivity than under-

Table 84. Summary of an Engineering Study of the Cost of Mining Coal in Western Kentucky

Production Million Tons per year	Operating Cost		Selling Price Dollars per Ton
	Dollars per Ton	Cents per Million BTU	
1	3.90	16.3	5.35
1 (2 seams)	2.98	12.4	3.81
3	2.58	10.8	3.46

Note: Selling price is the price per ton of coal needed to obtain a profit of 12 percent discounted cash flow.

Source: Bureau of Mines, "Cost Analyses of Model Mines for Strip Mining of Coal in the United States," *Information Circular* 8535, 1972.

ground operations because they typically use more machinery per man. Likewise, larger mines usually demonstrate higher productivity than small mines because they are likely to use more capital-intensive mining techniques. But a direct link between productivity of labor and cost per ton is at best a tenuous one. Without a sufficient data base, which currently is unavailable, this link cannot be verified empirically.

Although the evidence is not conclusive, the literature suggests that a large coal company with large financial resources would automatically select approximately optimal size of mine, select the best location for the mine, and use the most efficient technological combination of labor and capital equipment.

If this argument that adequate financial resources will permit a firm to attain an optimum position on all three points is correct, then one would expect large companies and their mines to be the lowest-cost producers of coal. Labor productivity would then follow the pattern shown by the data available. Thus, the evidence, though not conclusive, supports the argument that large mines can produce coal at lower cost than small mines.

Data published in engineering studies do not provide sufficient information and evidence on the costs of producing coal. Only one study is available for the region in question, and that study attempts to estimate the costs for only three surface mines. The estimates, which are shown in Table 84, suggest that some economies of large-scale production may be attainable, but this conclusion is quite tentative. Only more extensive empirical investigation could produce meaningful results. The link between cost of production and selling price remains unexplored and is largely a function of the com-

Table 85. Percentage Share of Output from Different Size Categories of Underground Mines, 1968-1973

	Year	Category* 1	2	3	4	5	6
Western Kentucky	68	90.97	3.34	2.04	1.78	1.78	0.07
	69	93.05	3.62	0.00	2.28	0.94	0.07
	70	93.30	1.12	3.72	1.22	0.53	0.07
	71	86.70	7.49	4.18	0.87	0.68	0.05
	72	93.92	3.24	1.31	0.92	0.47	0.13
	73	92.70	5.56	0.46	1.09	0.12	0.02
Indiana	68	86.48	0.00	5.30	3.41	4.79	0.00
	69	87.15	0.00	5.07	7.06	0.71	0.00
	70	91.16	0.00	0.00	7.16	1.28	0.38
	71	91.78	0.00	0.00	8.21	0.00	0.00
	72	65.70	21.16	7.19	5.88	0.00	0.00
	73	77.82	0.00	0.00	22.05	0.00	0.00
Illinois	68	92.61	4.24	1.92	0.56	0.64	0.01
	69	96.75	1.36	1.22	0.47	0.15	0.02
	70	96.69	2.25	0.73	0.19	0.11	0.00
	71	96.14	2.45	0.56	0.60	0.22	0.00
	72	98.86	0.00	0.95	0.00	0.20	0.00
	73	98.05	1.30	0.00	0.48	0.15	0.00

*Size categories are as follows, (annual output):
(1) 500,000 tons and over   (4) 50,000 to 100,000 tons
(2) 200,000 to 500,000 tons (5) 10,000 to 50,000 tons
(3) 100,000 to 200,000 tons (6) less than 10,000 tons

Sources: Calculated from Bureau of Mines, *Minerals Yearbook*, 1970-1973.

petitiveness of market forces and of the elasticities embedded in demand-and-supply relationships.

The third method, known as the survival technique, requires greater elaboration. If we assume that firms pursue the profit objective, or that meaningful competition exists, firms will gravitate over time toward that size of output at which production costs are lowest. If a particular size of operation generates unacceptably high costs of production, firms would tend to move to a different size operation in an effort to reduce their costs. Eventually, it can be argued, a firm's high-cost operations would tend to disappear in favor of lower-cost operations. Therefore, if one examines the size distribution of mines, over time, it would be expected that the share of output from the sizes of mines with lower operating costs would increase, while the share coming from mines with higher-cost operations would decrease. Tables 85 and 86 examine the share of coal output coming from the different sizes.

To understand fully how to interpret these tables, one must recall the

Table 86. Percentage Share of Output from Different Size Categories of Surface Mines, 1968-1973

	Year	Category* 1	2	3	4	5	6
Western Kentucky	68	91.82	5.87	0.37	0.92	0.94	0.05
	69	90.05	5.77	1.18	2.06	0.79	0.13
	70	88.40	5.68	1.78	1.22	2.18	0.14
	71	82.50	8.23	3.64	3.02	2.40	0.14
	72	89.18	4.72	2.21	2.21	1.47	0.20
	73	88.54	4.47	4.80	1.31	0.66	0.18
Indiana	68	86.60	10.13	0.82	1.08	1.21	0.14
	69	94.03	2.00	2.15	1.29	0.50	0.20
	70	96.92	0.00	0.00	1.85	1.03	0.18
	71	94.86	2.31	0.59	1.56	0.44	0.21
	72	94.57	1.18	1.11	2.07	0.98	0.11
	73	92.45	4.45	1.38	0.48	1.15	0.06
Illinois	68	91.63	7.00	0.64	0.44	0.25	0.02
	69	94.59	4.02	0.71	0.43	0.15	0.08
	70	97.51	1.32	0.56	0.25	0.28	0.05
	71	93.88	4.61	0.62	0.50	0.21	0.14
	72	94.76	2.90	1.43	0.73	0.17	0.02
	73	92.23	6.67	0.00	0.92	0.09	0.06

*Category sizes are the same as in Table 38.

Sources: Calculated from Bureau of Mines, *Minerals Yearbook*, 1970-1973.

nature of demand conditions during the period examined. From 1968 through 1970 output was expanding to meet a rapidly growing demand. The decline in output in 1971 resulted largely from a six-week industry strike, which appears to have affected some sizes of mines more than others. By 1974 western Kentucky's and Indiana's outputs had returned approximately to 1970 levels; Illinois output, on the other hand, had fluctuated considerably and in 1974 was actually slightly below the previous low of 1971.

In Table 85 an effort is made to relate relative changes in the share of output to size of operation. In underground mining, the table shows that between 1969 and 1973 the largest mines located in western Kentucky and Illinois increased their share of the market. The temporary decline for 1971 in western Kentucky probably is the result of larger mines being affected more severely by the labor strike than small mines. The reason for the dramatic change in the 1972 and 1973 size distribution for Indiana is unknown.

A large part of the growth shown in category 1 seems to have resulted from a decline in categories 3, 4, and 5, in both western Kentucky and Illinois. During the course of the 1971 strike, however, several mines in the above 500,000-tons-per-year category were forced to reduce operations. This pushed their output into size category 2 for that year.

Category 3 experienced substantial declines in both Illinois and Indiana, with the exception of 1972. In western Kentucky, output for this category was somewhat erratic, fluctuating from zero to 4 percent over the period examined. This would seem to indicate that category 3 is less of an equilibrium category and more of a transitory one, particularly when compared with mines that fall into size categories 1, 2, and 5.

Output for category 4 declined in western Kentucky and Illinois for the period while the fluctuations in Indiana remain unexplained. It should be noted, however, that practically no coal is mined with underground methods in Indiana; a mere 100,000 tons in 1974. Output for categories 5 and 6 declined steadily everywhere, supporting the view that small underground-mining operations are not economical in the Eastern Interior Coal Basin.

In summary, only the largest category of underground mines expanded during normal recent years, suggesting that this size of mine is economically better suited for low-cost production than the other categories. This is also true for the next largest category in western Kentucky. The other size classes tended to decline generally in relative importance. The evidence suggests that some economies of scale in underground mining may be obtainable in the basin.

For surface mining, Table 86 shows that the largest size categories in Indiana and Illinois have generally increased their share in output, mostly at the expense of the middle categories. The smaller operations (categories 5 and 6) have basically maintained their relative positions. This would suggest that there are some economies of scale to be found in surface mining in Indiana and Illinois at the higher spectrum of the mine-size scale: in categories 1 and 2. For western Kentucky, an opposite trend seems to emerge, for the larger categories declined, while the smaller ones remained unchanged during 1969-1973. This would suggest that few, if any, counties of large-scale operation exist in the larger-size operations in western Kentucky surface mining and that there may even be operative diseconomies of scale at that mine size. On the other hand, mines with annual surface output between 100,000 and 200,000 tons may offer a production cost advantage in this part of Kentucky.

Over the past eight years, surface mining in Indiana has steadily increased its share in total mining so as to render underground mining practically extinct. Table 87 shows that less than 1 percent of the state's 1974 output originated from underground mines.

In contrast, the share of surface mining in total western Kentucky and Illinois output has gradually declined to a point where in the case of the

Table 87. Percentage Distribution by Type of Mine, 1967-1974

	Year	Surface Mines	Underground Mines
Western Kentucky	67	65.35	34.65
	68	61.60	38.40
	69	58.22	41.78
	70	63.33	36.67
	71	66.83	33.17
	72	64.56	35.44
	73	58.53	41.62
	74	55.65	44.35
Indiana	67	91.26	8.74
	68	88.28	11.72
	69	89.50	10.50
	70	90.60	9.40
	71	91.76	8.24
	72	94.43	5.57
	73	96.88	3.12
	74	99.42	0.58
Illinois	67	57.10	42.90
	68	57.74	42.26
	69	53.53	46.47
	70	50.72	49.28
	71	49.59	50.41
	72	51.59	48.41
	73	47.10	52.90
	74	46.31	53.69

Sources: Calculated from Bureau of Mines, *Minerals Yearbook,* 1970-1973; "Coal—Bituminous and Lignite," *Mineral Industry Surveys,* 1973-1975.

latter, more than half of the state's output now originates from underground operations. In western Kentucky, surface operations still contributed 55 percent of the state's 1974 total, but this represents a 15 percent decline from 1967. Surface mining is becoming more costly as the constraint of the declining productivity of land is asserting itself.[12]

These statistics suggest the emergence of a changed cost-price relationship between surface and underground mining. In western Kentucky and Illinois, the gradual shift to more underground mining indicates that operators are beginning to find it more profitable to mine coal underground rather than at the surface. The most easily strippable coal has already been mined, whereas the most easily minable coal that lies underground has not. It can also be inferred, however, that Indiana operators perceive a distinct cost advantage in mining coal with surface methods.

There exist many factors which might have contributed to the apparent change in underground versus surface mining shares, such as particular kinds

Table 88. Average Railway Transportation Costs per Ton of Coal, U.S., 1967-1973

Year	Percent of Coal Output Shipped by Rail	Value per Ton (f.o.b. Mine)	Average Rail Freight Charge	Ratio of Rail Charge to Value (f.o.b. Mine)
1967	69.6	$4.62	$3.00	0.65
1968	67.5	4.67	3.01	0.64
1969	68.4	4.99	3.10	0.62
1970	67.1	6.26	3.41	0.54
1971	69.2	7.07	3.70	0.52
1972	66.2	7.66	3.67	0.48
1973	67.2	8.53	3.71	0.43

Sources: Calculated from Bureau of Mines, "Coal—Bituminous and Lignite," *Mineral Industry Surveys,* 1968-1974.

of coal mined, changing seam width and depth of overburden, local EPA restrictions, and mine safety standards. It lies beyond the scope of this survey to attempt to estimate the influence of each on the changed proportions of surface to underground mining.

In sum, the use of the three analytical techniques here described suggests that over the period studied, economies of scale associated with large mining operations have been small, with the notable exception of large western Kentucky surface mines. Unfortunately, no data are available to permit an answer to the question whether production costs of large mines are significantly lower than those of small mines. It would probably be correct, however, to assert that for particular seams in particular locations, small operations in surface and underground mining have production costs that differ very little, if at all, from those of large operations.

An important component of the delivered price of coal is transportation cost. It is regrettable that so little raw data are readily available on such an important aspect of this industry, but the scarcity of data is real and thus limits the discussion of this topic.

Some data concerning the nation as a whole are presented in Table 88. About two-thirds of all coal is shipped by rail. For the most recent years, the ratio of average rail freight charge to the FOB mine price per ton of coal has been declining steadily. The reason for this decline is the much more rapid

Table 89. Representative Rail Charges for Coal, May 1968

				Destination					
	Chicago, Illinois			Waukegan, Illinois			South Chicago		
Origin	Miles	Charge	Rate per Ton-Mile (Mills)	Miles	Charge	Rate per Ton-Mile (Mills)	Miles	Charge	Rate per Ton-Mile (Mills)
Western Kentucky	401	$2.53	6.3	472	$3.21	6.8	399	$2.95	7.4
Southern Illinois	317	2.25	7.1	394	2.79	7.1	321	2.58	8.0
Belleville	300	2.16	7.2	416	2.60	6.3	342	2.45	7.2
Springfield		not reported		365	3.36	9.2	192	2.21	11.5
DuQuoin		not reported		401	2.60	6.5	327	2.45	7.5

Source: Frank B. Fulkerson, "Transportation of Mineral Commodities on the Inland Waterways of the South-Central States," *Information Circular* 8431, 1969.

climb of coal prices than of rail transport costs; therefore, the significance of transport costs as a percentage of the delivered price of coal has diminished, and distant mining operations have become more competitive.

The structure of charges for rail transport is highly complex, and it is difficult to obtain clear and comparable data. Some publicly available data were found in a Bureau of Mines information circular, and, while certainly not complete, they are believed to be representative of the general patterns of rail freight charges. It should first be pointed out that in the calculations which follow, all coal originating in an area was assumed to bear the same charge for transport regardless of destination. Table 89 lists some representative freight charges in effect in May 1968. Two obvious points should be noted: the longer the distance the greater the total charge, but also, the longer the distance the lower the cost per ton-mile since the fixed costs can be spread over more miles.

A thorough study of the structure of freight rates was done by Reed Moyer in 1960 and led him to conclude that the rate advantage for the Midwest over the East rapidly disappears on shipments to destinations east of a north-south line drawn through eastern Indiana and southwestern Michigan, and that transportation costs included in the price of coal imported from outside the Eastern Interior Coal Basin are substantial enough to prevent Midwest coal users, some of the nation's largest consumers, from buying coal from producers not located in the region.[13] While the first conclusion is probably still valid today, the second is not. The costs of mining the newly explored surface lands in the Eastern Rocky Mountains are so low as to permit this coal, despite much higher transport costs, to compete successfully in the Midwestern markets with coal mined in the Interior Basin.

Table 90. Cost of Transporting Coal by Unit Train, 1969

Case	Location	Miles	Cost per Ton-Mile (Mills)
Inland Steel	Illinois	279	5.4
Captain Mine	Illinois	306	4.4
		354	5.2
Orient No. 5	Illinois	130	7.7
Bellcoal, Inc.	Kentucky	333	5.4
Bell and Zoller	Kentucky	129	7.8

Source: T. O. Glover, M. E. Hinkle, and H. L. Riley, "Unit Train Transportation of Coal: Technology and Description of Nine Representative Operations," *Information Circular* 8444, 1970.

Since rail charges are high and since rail is the most important form of transporting coal, it would seem that pressures would exist to reduce these costs. During recent years, efforts to reduce the costs of rail transport have led to the use of unit trains, a three-way arrangement among mine operator, coal purchaser, and the railroad company. Unit trains provide the most efficient use of the transportation facilities and thereby result in lower transportation costs. Table 90 summarizes case studies of unit trains reported by the Bureau of Mines. Cost per ton-mile of moving coal is appreciably lower for unit trains than for conventional trains. However, this cost break is available only to large operators because the arrangements typically call for minimum shipments of one million tons per year.

The second most important method of transporting coal is by river barge. Generally the cost per ton-mile is sufficiently lower for barge transport as compared to railways to offset the often additional miles of the circuitous route a barge must follow. The structure of rail rates does not favor western Kentucky coal. The large guaranteed market alluded to by Moyer is located in areas closer to Illinois and Indiana mines than to western Kentucky mines. However, Kentucky does hold an advantage in barge transportation. The Green River cuts through western Kentucky and connects the coalfields with the Ohio and Tennessee rivers, the Mississippi River, and eventually with the Gulf of Mexico.

On the other hand, Indiana has no navigable waterways, while Illinois at this time has only the Illinois waterway which runs north of the major mining areas. While this river advantage may not be sufficient to allow Kentucky to attain an equal share in the markets in Illinois and Indiana, it is vitally important in allowing Kentucky to compete in the South. Over 50 percent of the

Table 91. Methods of Transporting Indiana, Illinois, and Western Kentucky Mined Coal, 1973

	Western Kentucky		Illinois		Indiana	
	Thousand Tons	Percent	Thousand Tons	Percent	Thousand Tons	Percent
Total output	53,679	99.99*	61,572	99.91*	25,253	99.99*
Transported by						
Unit train	7,291	13.58	22,155	35.98	5,493	21.75
Other train	14,655	27.30	29,397	47.74	13,719	54.32
Total Trains	21,946	40.88	51,552	83.72	19,212	76.07
River Barge	21,638	40.31	2,159	3.51	1,917	7.59
Truck	1,365	2.54	3,392	5.51	3,682	14.58
Mine-mouth generating plants	8,707	16.22	4,399	7.14	411	1.63
Others	23	.04	70	.03	31	.12

*May not sum to 100 due to rounding.

Sources: Calculated from Bureau of Mines, "Coal—Bituminous and Lignite," *Mineral Industry Surveys,* 1974; and unpublished statistics.

total coal sold in the five southern states of Alabama, Mississippi, Georgia, Florida, and Tennessee is shipped by barge, and these states consume about 45 percent of western Kentucky's output. And even though overall Kentucky is at a transport disadvantage to markets located in the north, of the coal sold in Illinois, Minnesota, and Ohio, at least 90 percent is shipped via waterways.

Another method of transporting coal is by truck, the principal means of transportation in cases where distances are short. For states like Illinois and Indiana, whose coal is consumed mostly within their own boundaries, trucks are used a great deal. But trucks are not important for western Kentucky coal producers, since their markets are typically farther away.

Most recently, attempts have been made to minimize the transport of coal. Methods are being investigated to convert coal into electricity at or near the mine and to transport the electricity instead. All three states have in operation some mine-mouth generating plants, and as one might expect, western Kentucky with its transportation disadvantages to northern markets is most heavily involved in this kind of operation.

Table 91 summarizes the methods of transporting coal mined in western Kentucky, Illinois, and Indiana in 1973.

The data show that two-fifths of western Kentucky coal is shipped to its destination by barge. Only 3.5 percent of Illinois and 7.6 percent of Indiana coal is shipped by river. The transport cost advantage that western Kentucky operators have is very large and explains why they are able to compete in several very distant markets.

In an attempt to reduce rail-transport costs, mine operators in Indiana and Illinois have made strong efforts to integrate shipments of coal into unit trains. The success of these efforts has been proved by the fact that in 1973 more than one-third of Illinois coal and more than one-fifth of Indiana coal were transported in this way. The trend toward using more unit trains for transporting coal has been accelerating in recent years and is likely to continue into the future. In western Kentucky, this trend is not in evidence as increasingly more coal is being shipped by river.

Productivity is generally measured in terms of output per man-hour or man-day. Economic theory teaches that if other influences are disregarded, one can expect changes in productivity to lead to reciprocal changes in price. The link between productivity and price is established by production costs, so that, for example, higher productivity would mean lower production costs, which would mean lower prices.

As the statistics in Table 92 show, however, no such clear relationship between productivity and price can be observed in coal mining. For example, data for underground mining in western Kentucky show at least two years during which both output per man-day and price increased. Similar observations can be made about Indiana and Illinois. In Indiana, for example, output per man-day in underground mining in 1971 was little different from that of 1967, but prices had increased during that period by 46 percent. During 1972 productivity rose 8 percent, but price remained unchanged.

The price-productivity relationship in surface mining seems to follow a similar pattern. In western Kentucky, for instance, in 1968, 1970, and 1972, even though productivity increased over the previous year, price increased as well.

The statistics shown in Table 92 reflect short-run patterns only. Over the long run, however, one can say with some assurance that changes in productivity will influence changes in price, if not proportionately, at least partially. In other words, the recent declines in underground-mining productivity unquestionably will generate some cost and hence price increases in the future. When other factors influencing demand and supply are considered, however, such as changes in income, severe climatic upheavals, political and military conflicts, and oil-supply interruptions, then the productivity-price relationship becomes highly unpredictable, and all that can be said with certainty is that changes in productivity will have some implicit influence on price, but how much is indeterminable.

The data shown in Table 92 are plotted in Figure 11. The curves clearly

Table 92. Productivity and Price in Coal Mining, Eastern Interior Coal Basin, 1967-1973

	Underground				Surface[a]			
	Output per Man-day (Tons)	Percent Change	Price Per Ton (f.o.b. Mine)	Percent Change	Output per Man-day (Tons)	Percent Change	Price Per Ton (f.o.b. Mine)	Percent Change
Western Kentucky								
1967	22.63		$3.66		51.91		$3.30	
1968	23.83	+ 5.3	3.53	− 3.6	54.41	+ 4.8	3.36	+ 1.8
1969	25.64	+ 7.6	3.56	+ 0.8	48.22	−11.4	3.51	+ 4.5
1970	21.61	−15.8	4.73	+32.8	48.55	+ 0.7	3.92	+11.7
1971	18.45	−14.7	5.48	+15.8	46.68	− 3.9	4.50	+14.8
1972	17.02	− 7.8	5.97	+ 8.9	46.83	+ 0.3	4.81	+ 6.9
1973	18.32	+ 7.6	6.49	+ 8.7	45.01	− 3.9	5.53	+14.9
Indiana								
1967	16.72		$4.31		43.39		$3.87	
1968	20.40	+22.0	4.38	+ 1.6	37.03	−14.7	3.81	− 1.6
1969	17.73	−13.1	4.77	+ 8.9	40.56	+ 9.5	4.05	+ 6.3
1970	17.56	− 1.0	5.79	+21.4	40.51	− 0.2	4.47	+10.4
1971	16.02	− 8.8	6.61	+14.1	39.00	− 3.8	5.05	+12.9
1972	17.35	+ 8.3	6.62	+ 0.2	35.74	− 8.4	5.51	+ 9.1
1973	18.74	+ 8.0	6.94	+ 4.8	37.08	+ 3.8	6.04	+ 9.6
Illinois								
1967	22.38		$3.96		41.59		$3.83	
1968	22.17	− 1.0	4.14	+ 4.6	39.44	− 5.2	3.92	+ 2.3
1969	22.94	+ 3.5	4.43	+ 7.0	37.62	− 4.7	4.23	+ 7.9
1970	21.25	− 7.4	5.33	+20.3	33.58	−10.8	4.53	+ 7.1
1971	18.85	−11.3	5.96	+11.8	34.89	+ 3.9	4.95	+ 9.3
1972	17.87	− 5.2	6.83	+14.6	37.09	+ 6.3	5.49	+10.9
1973	18.07	+ 1.1	7.52	+10.1	35.80	− 3.5	5.81	+ 5.8

[a]Does not include auger mining.

Sources: Bureau of Mines, *Minerals Yearbook*, 1971; "Coal–Bituminous and Lignite," *Mineral Industry Surveys*, 1973-1974.

Figure 11. Relationship between Price and Productivity for the Eastern Interior Coal Basin by State, 1967-1973

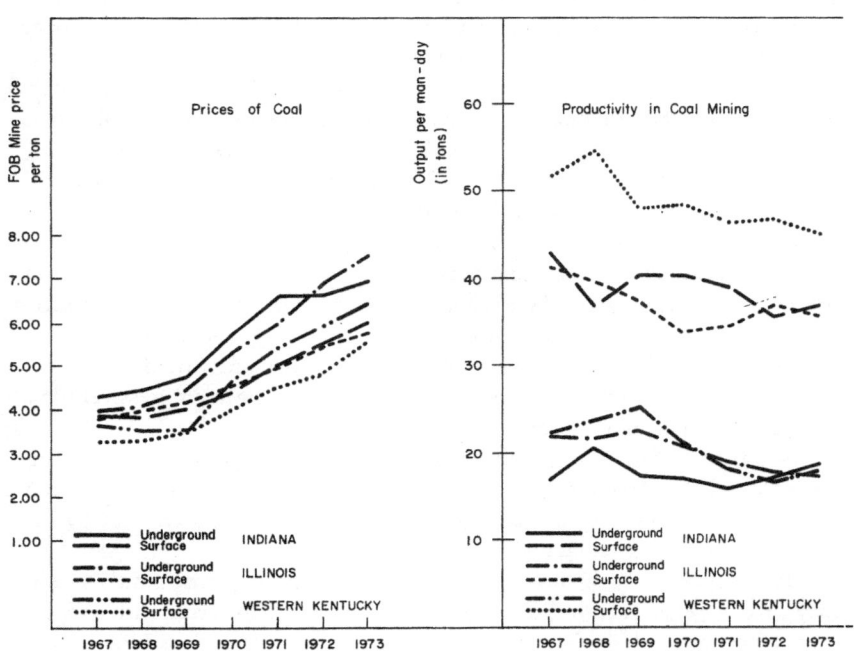

Source: *Minerals Yearbook,* 1967-1974.

show the substantially greater productivity of western Kentucky surface operations over that of neighboring states and the higher productivity of surface operations when compared with underground operations. There has been a general increase in underground-mining productivity in Indiana, but a remarkably uniform decline in western Kentucky and Illinois. The post 1969 general decline in underground productivity is the consequence of tightened health and safety standards, but the mines in Indiana seem to have adjusted to these more rapidly than those in either western Kentucky or Illinois where productivity reversed its yearly decline only during 1973.

Another interesting fact that emerges from Table 92 is that productivity throughout the basin was remarkably uniform in 1973. Output per underground laborer was over eighteen tons per day. Price levels, however, were more diverse, with Illinois coal as the most expensive, 16 percent higher than western Kentucky coal, the least costly. The explanation for this divergence in price probably lies in market structure and other institutional factors. Quality differences such as sulfur and ash content or BTU values could also explain the price difference.

Western Kentucky surface mines are considerably more productive than their counterparts in Indiana and Illinois. One would therefore expect western Kentucky prices also to be the lowest, which in fact they are. However, during the past four years they also increased by nearly 50 percent, whereas Indiana and Illinois strip-mined coal prices rose less rapidly. It is probable, therefore, that western Kentucky surface operators are obtaining significant economic rents which if reinvested in their operations could ensure continued favorable profit growth in the area.

A disquieting note emerges, however, from the continuing decline in western Kentucky strip-mining productivity. This decline is also evident in Illinois and Indiana with the exception of Indiana in 1973. One explanation for this general decline is the expansion of mining operations into less productive land. This is not all that surprising, since one would expect the most productive mineral lands to be mined first, with progressively less productive lands mined later.

Another explanation for the decline in strip-mining productivity is stricter enforcement of more stringent land-reclamation standards. This trend is likely to continue until reclamation efforts meet the standards set by the citizenry of the particular region.

Properly viewed, the prospects for western Kentucky coal and the industry must be examined in the context of the short and long run, with particular reference to the influence of noneconomic factors such as EPA restrictions, anti-strip-mining legislation, government-import policies, domestic fuel shortages, supply interruptions, and the price of foreign petroleum. The probability of occurrence and the possible impact of these factors, however, are at best uncertain. There prevails, for instance, at this time no unanimity of view on whether the country really is in the grips of a critical natural gas shortage or whether existing imbalances in demand and supply are merely artificial deficits created to dramatize the need for higher prices because industry profit margins are considered too low. The existence of a natural gas shortage could have a critical, but highly unpredictable, impact on the coal industry. This example illustrates that what follows, therefore, must necessarily be a tenuous assessment of the prospects for the western Kentucky coal industry, since much of its future welfare depends not only on supply-and-demand imbalances of other fuels but also on the fate of administration and congressional policies now under debate.

The southeastern United States, including Kentucky, which holds the major share of western Kentucky's coal market, since at least 1960, has been in a sustained period of economic growth. The impetus for this growth

Figure 12. Manufacturing Job Increases in the United States and the Southeastern Region, 1950-1970

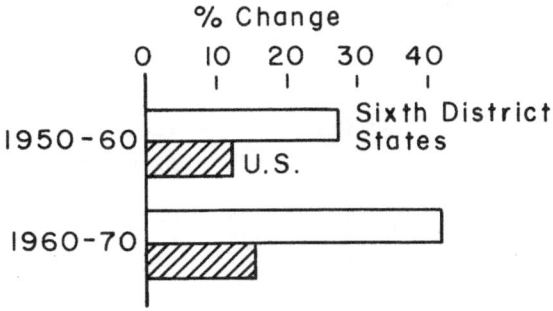

Source: Federal Reserve Bank of Atlanta, *Monthly Review* (January 1971).

derives mainly from the continued transition in the economic base of the area from an essentially agrarian to an industrial structure. This fact is clearly illustrated by the very rapid growth in manufacturing employment in the region as shown in Figure 12. During the 1950s and 1960s, manufacturing jobs increased much more rapidly in the Southeast than in the nation as a whole.

Further evidence of this growth can be found in changes in per-capita income for the area. Although still below the national average, per-capita income of the Southeast is rising to close the gap between it and the national level. Between 1968 and 1973, the six states' per-capita income climbed to 85 percent of the national average from 74 percent at the outset of the decade.[14] Table 93 shows the increase in per-capita personal income for the period 1968-1973. In every one of the southeastern states in which western Kentucky sells coal, including Kentucky itself, per-capita personal income increased more rapidly than the national average.

The state of Kentucky, which absorbs approximately 30 percent of western Kentucky's coal output, also has undergone significant economic growth in the 1960s, whether measured in terms of per-capita income or manufacturing employment.[15] The state has experienced a steady trend toward urbanization and industrialization accompanied by a rising demand for electric power. The growth of Kentucky's economy has been more rapid than that of the nation and indications are that this trend will continue into the 1980s.

In view of the recent growth experience in the Southeast and the development of an industrial and urban base in the region, continued economic

Table 93. Per Capita Personal Income, 1968 and 1973

	1968	1973	Percent Increase
United States	$3,436	$5,041	46.7
Southeast	2,731	4,282	56.8
Kentucky	2,666	4,033	51.3
Tennessee	2,634	4,095	55.5
Georgia	2,852	4,395	54.1
Florida	3,077	4,923	59.9
Alabama	2,429	3,871	59.4
Mississippi	2,185	3,556	62.7

Sources: *The 1973 World Almanac and Book of Facts* (New York: Newspaper Enterprise Association, 1972); *The 1975 World Almanac and Book of Facts* (New York: Newspaper Enterprise Association, 1974).

growth in the area is expected to exceed somewhat the national pattern. In the short run the outlook for continued prosperity is largely a function of the health of the economy. The demand for electric power, and therefore for coal, is dependent to a large extent on growth in the economy. In addition, consumer spending during 1976-1978 will probably be stronger than during 1974-1975. There is no reason to suspect that the Southeast will not participate strongly in the economic recovery now under way, and the demand for coal-fuel energy should continue to grow. In the long run, however, beyond the early 1980s, the outlook for the western Kentucky coal industry is not so easy to predict.

The major factor that governs the future welfare of the western Kentucky coal industry is whether the consuming states' ambient air-quality standards are implemented and enforced. If these standards (which are listed in Table 94 and summarized in Figure 13) remain unmodified and are ultimately enforced, given the current level of inplace sulfur-removal equipment, most of western Kentucky's coal markets would be adversely affected. Nine states in particular—Tennessee, Georgia, Alabama, Kentucky, Indiana, Illinois, Ohio, Michigan, and Wisconsin—would have to seek substitutes for high-sulfur western Kentucky coal, thereby contributing strongly to an already vigorous demand for low-sulfur coal.[16] Together these markets absorbed over 97 percent of western Kentucky's 1973 production. Natural gas cannot be considered a feasible alternative to western Kentucky coal, and oil can be used only for selected markets and under special conditions.

The magnitude of the potential scarcity of low-sulfur coal is illustrated in

Table 94. State Implementation Plan Sulfur Content Limitations from Regulations and Conversions, as Used in Supply/Demand Calculations

	Emission Limitations		Fuel Content Limitations			
	Pounds $SO_2/10^6$ BTU		Percent S by Weight		Equivalent in Pounds S/10^6 BTU	
	Oil	Coal	Oil	Coal	Oil	Coal
Alabama[P]	1.0-R	1.0-R	0.9-C	0.7-C	0.50	0.58
Alaska	1.0-R	1.0-R	0.9-C	0.5-C	0.50	0.42
Arizona[A]	1.5-R	1.5-R	1.3-C	0.7-C	0.72	0.58
Arkansas[A]	1.5-R	1.5-R	1.3-C	0.8-C	0.72	0.67
California[A]	1.5-R	1.5-R	0.5-C	0.7-C	0.28	0.58
Colorado	1.0-R	1.0-R	0.9-C	0.5-C	0.50	0.42
Connecticut[1]	0.55-R	0.55-R	0.5-R	0.5-R	0.28	0.42
Delaware	1.1-C	1.5-C	1.0-R	1.0-R	0.56	0.83
District of Columbia	0.55-C	0.75-C	0.5-R	0.5-R	0.28	0.42
Florida	0.8-R	1.2-R	0.7-C	0.8-C	0.38	0.67
Georgia	0.8-R	1.2-R	0.7-C	0.8-C	0.39	0.67
Hawaii	0.75-C	0.75-C	0.5-R	0.7-C[A]	0.28	0.58
Idaho[P]	0.25-C	1.5-C	0.3-R	0.7-R	0.17	0.58
Illinois	0.8-R	1.2-R	0.7-C	0.8-C	0.39	0.67
Indiana	1.2-R	1.2-R	1.1-C	0.7-C	0.61	0.58
Iowa	1.5-R	5.0-R	1.3-C	>3.0-C	0.72	2.50
Kansas	3.0-R	3.0-R	2.7-C	1.9-C	1.50	1.58
Kentucky[P]	0.8-R	1.2-R	0.7-C	0.7-C	0.39	0.58
Louisiana	3.0-R	3.0-R	2.7-C	2.0-C	1.50	1.67
Maine	1.7-C	2.1-C	1.5-R	1.5-R	0.83	1.25
Maryland	0.55-C	1.5-C	0.5-R	1.0-R	0.28	0.83
Massachusetts	0.55-C	0.45-C	0.5-R	0.3-R	0.28	0.25

145

Table 94 (continued)

	Emission Limitations		Fuel Content Limitations			
	Pounds $SO_2/10^6$ BTU		Percent S by Weight		Equivalent in Pounds $S/10^6$ BTU	
	Oil	Coal	Oil	Coal	Oil	Coal
Michigan	0.8-C	0.9-C	0.7-R	0.5-R	0.39	0.42
Minnesota	1.7-C	2.6-C	1.5-R	1.5-R	0.83	1.25
Mississippi	4.8-R	4.8-R	>3.0-C	>3.0-C	1.67	2.50
Missouri	2.2-C	3.4-C	2.0-R	2.0-R	1.11	1.67
Montana P	2.0-R	2.0-R	1.8-C	0.9-C	1.00	0.75
Nebraska A	1.5-C	1.5-C	1.3-C	0.8-C	0.72	0.67
Nevada	1.1-C	2.5-C	1.0-R	1.0-R	0.56	0.83
New Hampshire [1]	2.0-R	2.0-R	1.25-R	1.35-C	0.69	1.12
New Jersey	0.3-C	0.3-C	0.3-R	0.2-R	0.17	0.17
New Mexico	0.34-R	0.34-R	0.3-C	0.18-C	0.17	0.15
New York [2]	0.8-C	1.0-C	0.65-C	0.65-C	0.36	0.54
North Carolina	1.6-R	1.6-R	1.1-C	1.0-C	0.61	0.83
North Dakota	3.0-R	3.0-R	2.7-C	1.4-C	1.50	1.17
Ohio	1.0-R	1.0-R	0.7-C	0.55-C	0.39	0.46
Oklahoma	2.0-R	2.0-R	1.6-C	0.7-C	0.89	0.58
Oregon	2.8-C	2.1-C	2.5-R	1.0-R	1.39	0.83
Pennsylvania	1.8-R	1.8-R	1.6-C	1.2-C	0.89	1.00
Rhode Island	1.1-C	1.5-C	1.0-R	1.0-R	0.56	0.83
South Carolina	1.6-R	2.0-R	2.1-C	1.4-C	1.17	1.17
South Dakota	3.0-R	3.0-R	2.7-C	1.5-C	1.50	1.25
Tennessee	1.0-R	1.0-R	0.9-C	0.55-C	0.50	0.46
Texas	1.0-R	3.0-R	0.9-C	1.4-C	0.50	1.17
Utah	1.7-C	2.1-C	1.5-R	1.0-C	0.83	0.83
Vermont	1.0-C	1.5-C	1.0-R	1.0-R	0.56	0.83

Virginia	1.06-R	1.06-R	1.0-C	0.7-C	0.56	0.58
Washington	2.0-R	2.0-R	0.9-C	1.4-C	0.50	1.17
West Virginia[1]	2.7-R	2.7-R	1.5-R	2.0-R	0.83	1.67
Wisconsin	0.8-R	1.2-R	0.7-C	0.7-C	0.39	0.58
Wyoming[1]	1.5-R	1.5-R	0.5-R	0.5-R	0.28	0.42
Puerto Rico			1.0-R	1.0-R	0.50	0.83
Guam	2.81-R	2.81-R				
Samoa[1]	2.81-R	2.81-R	3.5-R	3.5-R	1.94	2.92
Virgin Islands			2.0-R	2.0-C	1.11	1.67

Key:

A   No regulations were available by 1 March 1972 and therefore Federal regulations were assumed.

C   Number was converted from units of the regulation.

P   The final state plan was not available 1 March 1972 and therefore the regulation was taken from the preliminary plan.

R   Number is in the units of the actual regulation.

[1]   The units of the regulation are inconsistant for conversion purposes.

[2]   In New York, the means of the state and New York City regulations are shown.

Source: Reproduced directly from The Mitre Corporation, *Impact of State Implementation Plans on Fossil Fuels.*

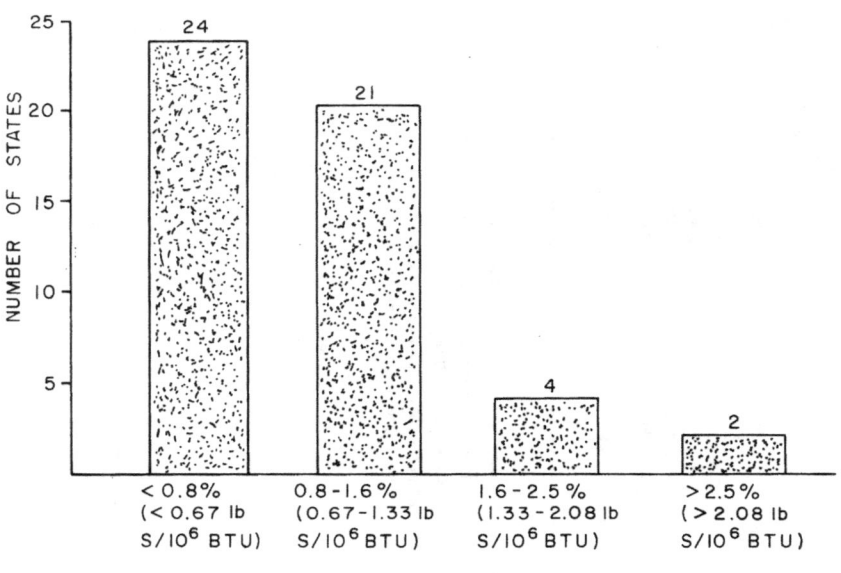

Figure 13. State Sulfur Regulations for Coal

Source: Reproduced from The Mitre Corporation, *Impact of State Implementation Plans on Fossil Fuels Availability and Requirements* (Washington, D.C., 1972).

Figure 14. The curves show the coal requirements and availabilities as a function of coal-sulfur content under State Implementation Plan (SIP) regulations. They are based on estimates summarized in Figure 15. These estimates include the use in the East of a moderate amount of Rocky Mountain coal, mostly from the eastern slopes. Even if a more generous estimate of the availability of low-sulfur coal is adopted, a deficit of at least 279 million tons in low-sulfur coal is forecast when the air standards are finally adopted.[17] The full impact of these scarcities will be delayed if the implementation of the SIP regulations themselves is deferred.

In summary, the prospects for the continued large-scale use of high-sulfur western Kentucky coal by electric utilities beyond the 1980s is uncertain. This prognosis is based at least in part on the assumption that sufficient equipment will not have been installed by then to control all sulfur emissions into the atmosphere. TVA, for example, has decided that it "would not be wise to undertake a massive investment in SO_2 [sulfur dioxide] removal systems of untested effectiveness."[18] Experimental programs, however, in particular the installation of limestone scrubbers to remove both particulates

Figure 14. Cumulative Coal Availability/Requirement at SIP Regulations, 1975

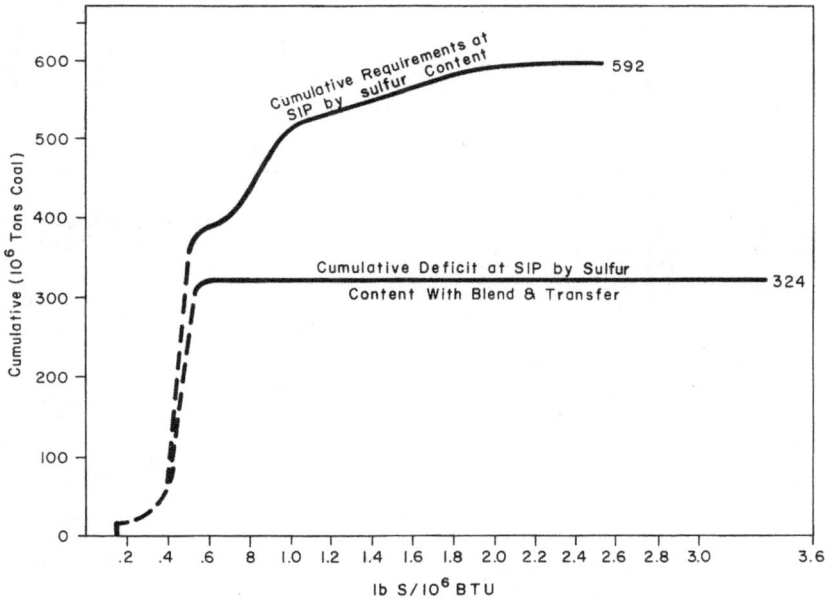

Source: Reproduced from The Mitre Corporation, *Impact of State Implementation Plans on Fossil Fuels Availability and Requirements* (Washington, D.C., 1972).

and sulfur dioxide from stacks, continue to be undertaken. But this system would not be an effective and economically feasible solution for controlling sulfur emission by existing plants. The cost of installation would be staggering. As a result, many public utilities are beginning to negotiate contracts for the purchase of low-sulfur coal.

In the case of power-generation equipment at new steam plants, the limestone scrubber method becomes a more attractive alternative because it can be incorporated into the original design. Of course, this would be done only after this method has been successfully tested. In the meantime, it appears that the most effective policy which TVA and other users of high-sulfur coal can pursue would be to obtain whatever sources of low-sulfur coal they can secure for future use and, also, to embark on a program of educating commercial and public users of electric power to conserve it.

In the long run all the additional costs associated with providing electric power without polluting the atmosphere will be passed on to consumers in the form of higher prices. Some of these additional costs may arise from the

Figure 15. Coal Availability/Requirements at SIP Regulations, 1975

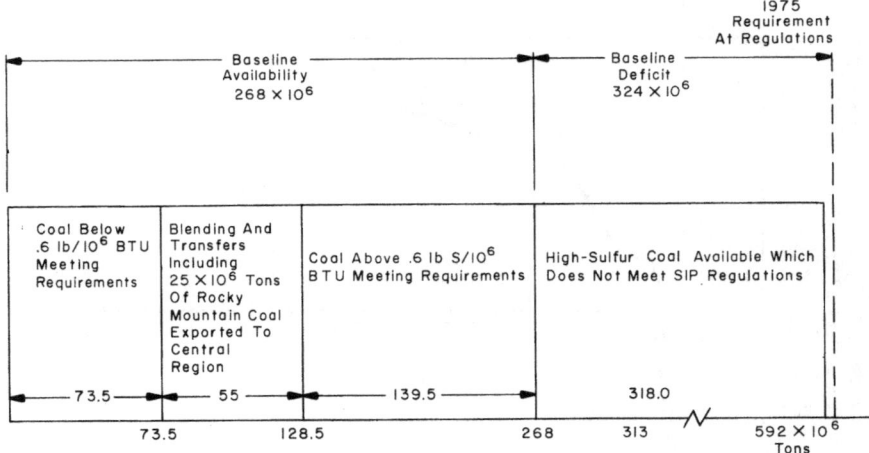

Source: Reproduced from The Mitre Corporation, *Impact of State Implementation Plans on Fossil Fuels Availability and Requirements* (Washington, D.C., 1972).

installation of equipment to cleanse coal prior to combustion; others, from efforts to remove sulfur from stack gases. Still other additional costs may arise from the need to continue purchasing more expensive low-sulfur coal. But whatever the source of these additional costs, higher electric-power prices have the effect of reducing the amount demanded at any one price and of stimulating the search for substitutes, both desirable results.

Considerable uncertainty surrounds the future of the western Kentucky coal industry today; an uncertainty fed by such institutional imponderables as the implementation of air-pollution standards and their enforcement, pending legislation against strip mining, fuel-import prices and policies, and foreign currency fluctuations. Any one of these issues individually can have a profound impact on the welfare of the industry, and collectively they can spell its continued growth or eventual decline. It would be futile at this point to speculate on which of these issues might be resolved, when and in what way, and on the probable impact of the remaining ones on the industry. But it is important to recognize that such issues are frequently decided less on the basis of economic and technical calculus, and more on the basis of political considerations, most of which lie beyond the scope of this study.

4. The Future Use of Energy Resources

Current technology makes possible the use of four major types of energy resources: coal, petroleum, natural gas, and uranium 235. In the future, others such as shale oil, uranium 238, and even solar energy may join the group, but for the present the former head the list. Converting energy resources into BTU equivalents, an accepted though not altogether meaningful practice, points to coal as a major energy resource available in the near term.[1] Table 95 shows the four major energy resources by location and type of fuel. Of the four fuels, only petroleum, coal, and natural gas are obtainable at present in sizable amounts. Shale-oil recovery is in an uncertain state of development and nuclear power is the target of continuing controversy.

The enormity of United States coal reserves, when contrasted with other fossil fuel reserves, except shale oil, is staggering. But whether these reserves actually will be developed in large quantities is primarily a function of the availability and price of petroleum.

The past few years have shown that for limited time periods the welfare of the coal industry closely parallels fluctuations in the price of petroleum. That is, the price of coal parallels, albeit at a lower level, the price of petroleum, principally residual fuels.[2] This so-called price tracking means that when the price of oil rises, coal becomes a more attractive alternative, where technically feasible. To the extent to which this takes place, demand pressures tend to raise the price of coal. A most vivid illustration of this price relationship is found in the spiraling price of coal in 1973 and 1974. Price increases in coal for the three years preceding 1973 were largely the consequence of a rising demand for low-sulfur coal and of new mine-safety legislation.[3] But beyond 1973, the price of coal, in particular the spot-market price, rose principally because embargo restrictions reduced sharply available oil supplies and because the price of imported oil went up fivefold. Uncertainty regarding the future availability of adequate supplies of imported oil plus its sharply higher prices, generated a strong shift in demand for the most suitable substitute—coal. Its price rose precipitously.

Whether the price of petroleum will continue to escalate and thereby have a continuing salutary impact on the coal industry is difficult to forecast.

Table 95. Recoverable Energy Resources by Type of Fuel and World Region, 1970
(quadrillion [10^{15}] BTU)

Fuel	United States	Western Europe	Persian Gulf North Africa	Rest of World	Total
Petroleum					
Proven Reserves	213	70	2,543	756	3,582
Unproven but recoverable	350	34	1,755	2,103	4,242
Coal	33,588	8,626	0	17,915	60,129
Natural Gas	447	83	3,409	2,268	6,207
Those presently available	34,598	8,813	7,707	23,042	74,160
Shale oil					
(Not presently available)	11,362	1,090	0	12,328	24,780
Total Fossil	45,960	9,903	7,707	35,370	98,940
Nuclear					
U-234					1,504,100
U-238					206,970,000
Total Nuclear					208,474,100

Source: W. D. Nordhaus, "The Allocation of Energy Resources," *Brookings Paper on Economic Activity* 3 (Washington, D.C., 1973), p. 542.

Recent events seem to support the view that in the foreseeable future, price increases in petroleum will occur with some regularity, but probably will be moderate.[4] During this period, other petroleum-exporting nations, notably China, can be expected to enter the world market on a large scale. A recent study concludes that by 1980, China will be in a position to produce roughly 8 million barrels of oil per day with a significant portion exported.[5] The long-term inference that can be drawn from this prediction is that, if correct, the monopolistic license acquired by the OPEC nations may become seriously circumscribed in the 1980s. It is unreasonable to expect that the People's Republic of China will join OPEC. A more likely prospect would be for it to export its surplus oil at governing world prices. But whatever the strategy, the addition of Chinese petroleum to world supplies will exert a downward pressure on its price or on efforts to raise it.

In addition to the prospect of China's becoming a major oil-exporting nation, the durability of OPEC as a cartel may also be questioned. The lessening of international tensions has eroded the ideological unity on which OPEC was initially founded. And on economic grounds cartels are traditionally unstable. If one large or several small members break away, a stampede for the exit can easily ensue. The only economic reason for the existence of a cartel is to counteract the potential dangers to profits from excess capacity. If production can be expanded at costs below price, competitive pricing and a threat to profits become very real. In order to counteract such a risk, a cartel is formed. However, with the depletion of petroleum reserves in the Persian Gulf area proceeding at only 1.5 percent per year, a rather uneconomical inventory depletion rate, and considerable excess capacity, the potential rewards for successfully circumventing the OPEC price become attractive indeed.[6]

Recognizing the vulnerability of the coal industry to the potential of falling oil prices, as unlikely as this prospect may appear today, the task force of Project Independence recommends measures to guarantee the future market for coal.[7] A variety of schemes is considered, but they would extract a heavy cost from the ultimate user of products or services produced with coal. The United States economy today, and in particular the energy industry, is firmly structured around the use of oil.[8] Little can be done in the short run to alter the fact that oil and gas account for more than three-fourths of the nation's energy consumption. This circumstance is, of course, not surprising when the versatility of oil is considered along with the relative ease with which sulfur can be removed from it. When contrasted with coal, petroleum's virtues are many.

Proposals for more energy self-sufficiency and ultimately energy independence, which would result in a vastly greater dependence on coal, would be very costly to implement. Even if such a policy is seriously pursued, which seems less likely today than in 1974, the ability of the nation to use its coal is limited. In the first place, coal cannot be used to propel automobiles, planes, and trucks, and second, for the next decade the capacity to burn coal to produce electric power is relatively fixed.[9] An independent evaluation of the costs of energy self-sufficiency, of satisfying national needs from domestic sources, puts the annual bill at a conservative $16 billion.[10] This would be the cost to the American economy of not engaging in free unrestrained energy trade devoid of oil tariffs, quota restrictions, or domestic energy subsidies.

Ultimately, the long-term prospects of the United States coal industry will be determined by government policy at home and abroad. As desirable as it may be from an economic standpoint to pursue a policy of free trade in energy resources, the politics of national interest will probably dictate otherwise. As William D. Nordhaus observes, "Oil mixed with politics is a volatile brew."[11] In the short run, the next five to ten years, the coal industry can be expected to grow moderately. Given the structure of the industry and the ownership pattern of the major coal reserves, it is unlikely that dramatic increases in output will develop. But increases in demand will continue to keep prices and profits at levels above historic averages. The long run is much more difficult to predict. However, it is reasonable to expect that research into new energy resources, for example solar energy, will begin to bear fruit and that this new energy source, along with a general emphasis on conservation and a continued reliance on domestic and imported petroleum, will dominate the world energy supply. Coal will probably not be able to expand its share in the world market significantly.

Current projections suggest that the United States will at least double its annual power consumption by the year 2000. To meet this increased demand and yet stay within ecological constraints, substantial changes in the methods of power generation will be necessary. To attempt to predict what form these changes may take would be far beyond the scope of this book. All that can be said, at this time, is that future patterns in power generation ultimately will be the result of a host of new economic, political, and technical considerations.

A controlling factor in converting existing methods of power generation to new techniques or to develop entirely new concepts is time. Most current proposals can best be divided into those which could be implemented in the foreseeable future, perhaps by the end of the century, and those which over a longer period of time can be expected ultimately to produce a balanced solution.

Some analysts point out that under current technology the only fuel available to meet anticipated energy requirements in the near future is coal. Table 96 shows that domestic oil and natural gas do not have the necessary reserves to provide a continuous source of power in the future. This fact has convinced some people that coal will be the fuel of the future, even to the point where coal will be used to produce oil and natural gas—presently coal's principal competitive energy sources. Just such a situation was described by H. M. Dole, assistant secretary of interior for mineral resources. He envisions the establishment of totally integrated plants processing daily as much as 100,000 tons of coal, 100,000 barrels of oil, 25,000 tons of solvent-refined coal, 400 million cubic feet of pipeline gas, a small amount of other chemicals, and 35 million kilowatts of electricity. Although such a plant might require $750 million to $1 billion of initial capital investment, the project would be economically feasible according to Dole.

While current technology favors the continued use of coal as an energy source, new technological developments have convinced other authorities that atomic power will be our most important power source in the future. They predict that the successful development of the breeder reactor with its unlimited nuclear reserves would dwarf alternative energy sources, including coal, and would provide power at relatively low cost.

Table 96. Estimated Recoverable Reserves, U.S.
(bituminous coal equivalents, millions of tons)

Bituminous coal	671,049
Crude oil	6,560
Natural gas	10,826
Natural gas liquids	1,247
Total	689,682

Note: Date of estimate for crude oil, natural gas, and natural gas liquids is January 1, 1970. The date for the coal estimate is January 1, 1967.

Source: National Coal Association, *Bituminous Coal Facts,* 1971, pp. 93-94.

Even though technology to obtain power from both of these sources is available today, the question of what will happen in the future remains unresolved. Ecologically motivated complaints against coal are strong, and recently enacted legislation, as well as proposed legislation, could increase the cost of coal substantially and thereby limit its future use. Nuclear power, on the other hand, has developed very slowly in light of the original predictions made in the 1950s and updated in the 1960s. Unless faster progress is recorded in the next twenty-five years than was attained in the last twenty-five, nuclear power's contribution may remain just a promise.

Society today is producing power by harnessing exhaustible resources. In time, even with successful breeder reactors, the supply of dependable fuels will run out. Already this fact has technicians looking for potentially new and continuous supplies of energy. These include such concepts as tidal energy, geothermal energy, and solar energy. Of these three, only solar power is considered a real alternative because it is estimated that the combined potential output from geothermal and tidal sources would produce only about 2 percent of the energy needs for the year 2000 (see Table 97).

In contrast, solar energy would be the ultimate ecologically balanced source of power. Under current technology, however, extensive capital investments would be required to develop and implement techniques to harness solar energy on a broad scale. Some experts therefore contend that so long as traditional energy resources are still available at reasonable prices, solar and other new energy sources will remain underdeveloped. Others, on the other hand, express great hope in the early use of solar energy for the heating and cooling of buildings. Within the next decade, and perhaps sooner if sufficient incentives exist for rapid implementation, solar heating and cooling can begin to contribute significantly to our energy supply.[12] In that event, the demand for coal derived from the general demand for electric

Table 97. Estimates of Supply of Continuous (Renewable) Energy (10^{12} Watts) for the U.S.

Source	Maximum Potential	Potential by 2000
Solar radiation	1,600.0	0
Tidal energy	0.1	0.06
Geothermal	0.01	0.006
Others	1.1	0.2
Total	1,601.21	0.266
Annual demand, year 2000 (approximate)		5-6

Source: Chauncey Starr, "Energy and Power," *Scientific American* 225 (September 1971): 43.

power to heat and cool residences and buildings could begin to undergo a marked moderation.

The principal economic obstacle to the use of solar energy is its high initial cost. In terms of lifetime energy costs, however, which includes low operating costs, the cost of using solar energy is moderate. Still, the need for high initial investment-capital to install a solar system represents a sizable institutional barrier.

Selection of the principal source of power will depend in part on the rate of technical development, in part on the economics of costs, but principally on political decisions. The political process undoubtedly will dominate future choices. Legislation pertaining to EPA restrictions or the Mine Health and Safety Act are examples not of economics or engineering but of political decision-making. These are basically political solutions. So is legislation concerning strip mining, taxation, coal transport, and other matters, all of which could easily increase the cost of coal relative to other fuels, in particular oil, so as to reduce its large-scale future use.

An interesting, yet generally unresearched, question that arises from the recent price escalations of energy resources, both oil and coal, is what long-term consequences these sharp price increases will have on the demand for capital equipment and labor. The results from a recently published study provide some noteworthy insights into this issue. They show that in the long run a rising price of energy will reduce the demand for it, that is, that the demand for energy is moderately price-elastic.[13] The answer to the question of what can be substituted for more expensive energy inputs, however, is unclear. Labor is only slightly substitutable. That is to say, the constraints of our technological structure are such as to make the substitution of labor, or any other nonenergy input, for energy unlikely. But energy and real capital were found to be complementary, which implies that as the price of energy rises and less of it is used, the demand for real capital will decline. In the short run, industry adjustment to rapidly rising energy prices is difficult; therefore, per-unit costs of production rise. In concentrated markets, some of these higher costs may be recouped by passing them on to the consumer in the form of higher prices. In competitive markets, however, particularly where large numbers of substitutable consumer goods or services exist, passing the higher costs on may not be possible, and profits will decline.

In the long run, since few nonenergy substitutes are available, some modification in the underlying technical structure of the economy away from energy intensive output can be expected. Traditional policies such as investment tax credits, accelerated depreciation, and others, work in the direction of reducing capital-service costs including the entire cost of employing an energy intensive production facility. Demand for capital and for energy therefore increases. To the extent to which this occurs, these government policies interact at cross-purposes with pricing policies designed to discourage and to conserve the use of energy. In short, sharply rising prices of oil and coal raise the costs of energy, which in turn discourage the use of new capital and energy-intensive technology at the expense of gains in productivity. The pace of growth in real output of goods and services is thereby reduced.

Notes

Chapter 1

1. Some notable exceptions are Richard L. Gordon, *U.S. Coal and the Electric Power Industry* (Baltimore, Md.: Johns Hopkins University Press, 1975); R. Moyer, *Competition in the Midwestern Coal Industry* (Cambridge, Mass.: Harvard University Press, 1964); J. Henderson, *The Efficiency of the Coal Industry* (Cambridge, Mass.: Harvard University Press, 1958).
2. Federal Reserve System, Board of Governors, *Federal Reserve Bulletin* (September 1974), p. 616.
3. Carl Kaysen and Donald Turner, *Antitrust Policy* (Cambridge, Mass.: Harvard University Press, 1959).
4. S. N. Whitney, *Antitrust Policies* (New York: Twentieth-Century Fund, 1958), 1:386.
5. James V. Koch, *Industrial Organization and Prices* (Englewood Cliffs, N.J.: Prentice-Hall, 1974), p. 78.
6. Reed Moyer, "Competition and Performance in the Midwestern Coal Industry," *Journal of Industrial Economics* (July 1967): 242.
7. Charles Rivers Associates, *The Economic Impact of Public Policy on the Appalachian Coal Industry and the Regional Economy*, prepared for Appalachian Regional Commission (Washington, D.C.: U.S. Department of Commerce, 1973), p. 223.
8. In May of 1974 the spot-market price was four to five times that of long-term contracts; *Lexington Herald-Leader*, May 26, 1974.
9. For a specific example, see testimony presented by the Tennessee Valley Public Power Association, the American Public Power Association, and the National Rural Electric Co-op Association before the Subcommittee on Small Business, 92d Congress, 1st session, July 1972. Statements and testimony are published in "Artificial Restraints on Basic Energy Sources," *Concentration by Competing Raw Fuel Industries in the Energy Markets and Its Impact on Small Business* (Washington, D.C.: Government Printing Office, 1973), 1:A445.
10. Moyer, *Competition in the Midwestern Coal Industry*, p. 38.
11. Ibid., p. 46.

Chapter 2

1. United States Federal Trade Commission, *In the Matter of Kennecott Copper Corp., Finding of Fact, Conclusion and Final Order*, Docket 8765 (Washington, D.C., May 5, 1971), p. 5.
2. Calculated from data published in U.S. Department of the Interior, News Release, March 10, 1974.

3. The Environmental Protection Agency standards scheduled to take effect in the next several years will require coal-burning electric utility plants without sulfur dioxide removal equipment to use coal with no more than 0.7 percent sulfur. Roughly 46 percent of eastern Kentucky reserves meet this standard, and another 28 percent contain only 0.8 percent to 1.0 percent sulfur. Source: National Coal Association, *Bituminous Coal Facts,* 1972 (Washington, D.C.: National Coal Association, 1973), p. 75.

4. About 22 percent of U.S. coal shipments are in these categories, while the corresponding figures for eastern Kentucky and the rest of District 8 (of which eastern Kentucky is a part) were 29 percent and 41 percent, respectively, in 1972.

5. Charles River Associates, *Economic Impact of Public Policy,* 1:98.

6. U.S. Department of Commerce, *Historical Statistics of the U.S., Colonial Times to 1957,* and *Statistical Abstract of the United States* (Washington, D.C.: Government Printing Office, 1973).

7. Real GNP is a measure of the nation's total output of final goods and services, valued at the prices of some base year. Changes in real GNP thus reflect changes in actual quantities produced. Changes in current-dollar GNP (valued at current prices) may reflect quantity and/or price change. Since 1900, real GNP has gone up tenfold to elevenfold, while electric generation is up three-hundred-fold. Electricity production has gone up twice as fast as real GNP since 1950. Ibid.

8. The use of electricity to perform some function which could also be performed through the direct application of fuels multiplies the direct energy input beyond that which would have been required with the direct fuel use.

9. The cost to a customer with a monthly usage of 500 kilowatt-hours remained constant from 1940 to 1969 at 2.1 cents per kilowatt-hour (Federal Power Commission, "Typical Electric Bills," *Annual Reports*). During that period, consumer prices more than doubled (Bureau of Labor Statistics, *Monthly Labor Review,* various issues).

10. Edison Electric Institute, *Fuels for the Electric Utility Industry, 1971-1985* (New York: Edison Electric Institute, 1973), p. 24.

11. On this point, see Edward W. Erickson and Robert M. Spann, "Supply Response in a Regulated Industry: The Case of Natural Gas," *Bell Journal of Economics and Management Science* 2 (Spring 1971): 94-121.

12. *Wall Street Journal,* April 5, 1974.

13. Battelle Memorial Institute, *Energy Perspectives* (Columbus, Ohio, 1973), p. 3.

14. Ibid.

15. " 'Breeder' Nuclear Plants Cleared in Study by AEC," *Wall Street Journal,* March 15, 1974.

16. Ibid.

17. Coal-distribution statistics for eastern Kentucky are counted as part of District 8, which includes, in addition to eastern Kentucky, 15 counties in West Virginia, 7 counties in Virginia, 9 counties in Tennessee, and all of North Carolina. Since information is published only for the entire district, statistics relating specifically to eastern Kentucky are available only for a few selected years.

It must be pointed out that the Bureau of Mines estimates the data on the basis of partial samples of the region's mines and that a margin for error must, therefore, be allowed. Bureau of Mines officials have indicated that they believe the data to be correct, having rechecked them for computational accuracy. Projections made in this study rely on these data but should be viewed as tentative.

18. C. E. Harvey, *Enforcing Weight Restrictions on Eastern Kentucky Roads* (Lexington: Office of Development Services and Business Research, University of Kentucky, 1972).

19. Charles River Associates, *Economic Impact of Public Policy.*

20. These forecasts disregard the possibility of any coal sales by eastern Kentucky mines to electric utilities west of the Mississippi River. Actually, 790,000 tons of coal were sold in 1972 to utilities in Iowa and Minnesota. The data indicate virtually no prior volume of shipments to these states from District 8, however. It is assumed that 1972 was not the forerunner of increased sales in this region.

21. This section draws heavily upon the analysis of the demand for coking coal in the Charles River Associates study cited above.

22. Imports as a fraction of domestic steel shipments rose from 11 percent in 1965 to 21 percent in 1971, but fell to 19 percent in 1972 (American Iron and Steel Institute, *Annual Statistical Report,* various issues).

23. Bureau of Mines, *Minerals Yearbook,* 1972.

24. Charles River Associates, *Economic Impact of Public Policy,* 1:191.

25. "Only about 50 percent of underground and 90 percent of the surface minable coal can be physically recovered by conventional mining methods." Federal Energy Administration, *Project Independence* (Washington, D.C.: Government Printing Office, 1974), p. 103.

26. U.S., Congress, Senate, Council on Environmental Quality, *Coal Surface Mining and Reclamation* (Washington, D.C.: Government Printing Office, 1973), p. 7.

27. Michael Rieber, *Low Sulfur Coal: A Revision of Reserve and Supply Estimates* (Urbana: Center for Advanced Computation, University of Illinois, November 30, 1973).

28. Ibid., p. 1.

29. This estimate was prepared by excluding overseas exports and captive output from production originating in Pennsylvania, West Virginia, Ohio, and eastern Kentucky and by assuming that 80 percent of the remainder is sold on long-term contract.

30. A news story in the *Lexington Herald-Leader* (May 26, 1974) noted that at that time spot-market prices were four to five times higher than the price of long-term contracts.

31. Eastern Kentucky is divided into four coal districts: Harlan, Hazard, Martin, and Pikeville.

32. Average size tells us little about the distribution of mine sizes around this average in the various regions.

33. Mathematica, Inc., *Design of Surface Mining Systems in Eastern Kentucky,* prepared for Appalachian Regional Commission, Department of Natural Resources and Environmental Protection, Frankfort, Ky., January 1974, Section 5.

34. *An Analysis of Strip Mining Methods and Equipment Selection,* prepared for the Office of Coal Research, U.S. Department of Interior, May 1973, pp. 61-63.

35. C. E. Harvey, *Enforcing Weight Restrictions on Eastern Kentucky Roads.*

36. "The decline in employee productivity is almost entirely due to the fact that nonproducing workers have had to be hired," S. Schweitzer, *The Limits to Kentucky Coal Output: A Short Term Analysis* (Lexington, Ky.: Institute for Mining and Minerals Research, 1973), p. 7.

37. See Mathematica, Inc., *Design of Surface Mining Systems in Eastern Kentucky.*

38. See M. E. Greenbaum, *Kentucky Coal Reserves: Effects on Coal Industry*

Structure and Output (Lexington, Ky.: Institute for Mining and Minerals Research, 1975).

39. Mary Jean Bowman and W. Warren Haynes, *Resources and People in East Kentucky: Resources for the Future* (Baltimore, Md.: Johns Hopkins Press, 1963), pp. 364-65.

40. Richard L. Gordon, *U.S. Coal and the Electric Power Industry*, p. 12.

41. Nine of the ten largest steel companies are located within easy reach of eastern Kentucky mines, and in the case of at least three, eastern Kentucky appears to have a decided transport advantage over West Virginia and Pennsylvania mines. Only Kaiser Steel is located west of the Mississippi River.

42. A promising new method being developed by the Illinois Institute of Technology involves the injection of chemicals into the exhaust stack and combining them with the sulfur oxide gas emissions from the burning coal. The resulting product is a liquid fertilizer that can be used in liquid or granulated form.

Chapter 3

1. U.S. Department of the Interior, Bureau of Mines, *Mineral Industry Surveys*, April 9, 1976; and Kentucky Department of Human Resources, Division of Research and Special Projects, unpublished data.

2. U.S. Department of the Interior, Bureau of Mines, *Analysis of Tipple and Delivered Samples of Coal*, Report of Investigation 7346 (Washington, D.C.: Bureau of Mines, 1969).

3. Darrell Gilliam and David Whitehead, *Employment and Expenditures Effect on the Kentucky Coal Industry in 1970* (Lexington, Ky., 1972), p. 32.

4. See U.S. Department of the Interior, Bureau of Mines, "Demonstrated Coal Reserve Base of the United States on January 1, 1974," *Mineral Industry Surveys* (Washington, D.C.: Bureau of Mines, June 1, 1974); and Tennessee Valley Authority, *Coal Reserves of Western Kentucky* (Chattanooga: Tennessee Valley Authority, 1969).

5. U.S. Department of the Interior, Bureau of Mines, "Demonstrated Coal Reserve Base of the United States by Sulfur Category on January 1, 1974," *Mineral Industry Surveys* (Washington, D.C.: Bureau of Mines, May 1975).

6. Federal Power Commission, *1970 National Power Survey* (Washington, D.C.: Government Printing Office, 1971), 1:1-13.

7. U.S., Congress, House, Committee on Science and Astronautics, Subcommittee on Energy, *Energy Facts* (Washington, D.C.: Government Printing Office, November 1973), p. 33.

8. Remarks by J. H. Wagner, Chairman, *TVA News Release,* January 24, 1973.

9. A term used throughout this study is "economies of scale." This term describes a situation where the average per-unit cost of producing a commodity is lower for large output levels than for small output levels. When there are no obtainable economies of scale, average per-unit costs for all sizes of output are similar. When average per-unit costs increase with size of output, diseconomies of scale are said to prevail.

10. This is not meant to be an estimate of actual remaining life. It is simply the reciprocal of the percentage of estimated reserves mined in the latest year.

11. M. E. Greenbaum, *Kentucky Coal Reserves*, p. 18.

12. M. E. Greenbaum, *Kentucky Coal Reserves*.

13. Reed Moyer, *Competition in the Midwestern Coal Industry.*
14. The six states of the region are Alabama, Florida, Georgia, Louisiana, Mississippi, and Tennessee.
15. Richard D. Carter and Janet S. Dinkel, "Economic Growth and Change in Kentucky, 1960-1970," *Economic Review* (Cleveland: Federal Reserve Bank, Oct.-Nov. 1972).
16. The Mitre Corporation, *Impact of State Implementation Plans on Fossil Fuels Available and Requirements* (Washington, D.C.: Mitre Corp., August 1972), p. 50.
17. Ibid.
18. Remarks by Mr. A. J. Wagner, chairman, *TVA News Release*, January 24, 1973.

Chapter 4

1. BTU = British thermal unit. The conversion factors used are: 5.8 million BTU/barrel of petroleum; 25.8 BTU/ton of bituminous coal; 1,000 BTU/cubic foot of natural gas.
2. *Coal Week*, September 15, 1975.
3. W. D. Nordhaus, "The Allocation of Energy Resources," *Brookings Papers on Economic Activity* 3 (Washington, D.C.: Brookings Institution, 1973), pp. 560-61.
4. The member nations of the OPEC countries have announced their intention to raise the price of oil in regular intervals. OPEC meeting, Vienna, Austria, September 1975.
5. Bobby A. Williams, "The Chinese Petroleum Industry: Growth and Prospects," in *China: A Reassessment of the Economy,* Joint Economic Committee, Congress of the United States (Washington, D.C.: Government Printing Office, 1975).
6. Federal Energy Administration, *Project Independence: Coal* (Washington, D.C.: Government Printing Office, November 1974).
7. Allen Hammond et al., *Energy and the Future* (Washington, D.C.: American Association for the Advancement of Science, 1973), p. 3.
8. M. A. Adelman, "Is the Oil Shortage Real?" in *Economics of Energy*, ed. Leslie E. Grayson (Princeton, N.J.: Darwin Press, 1975).
9. Morris A. Adelman et al., "Energy Self-Sufficiency: An Economic Evaluation," *Technology Review* (May 1974).
10. William D. Nordhaus, *The Allocation of Energy Resources,* p. 567.
11. Ibid., p. 564.
12. A recent Energy Research and Development Administration (ERDA) study estimates that in the year 2000, solar heating and cooling could have an energy impact equal to 5.9 quads, which is equal to approximately 250 million tons of bituminous coal or 1,062 million barrels of petroleum. ERDA, *Creating Energy Choices for the Future* (Washington, D.C.: Government Printing Office, 1975), vol. 1.
13. E. R. Berndt and David O. Woods, "Technology, Prices and the Derived Demand for Energy," *Review of Economics and Statistics* 57 (August 1975).

Bibliography

Adelman, Morris A., et al. "Energy Self-Sufficiency: An Economic Evaluation," *Technology Review,* May 1974.

American Academy of Political and Social Science. *The Annals. The Energy Crisis: Reality or Myth.* Philadelphia, 1973.

An Analysis of Strip Mining Methods and Equipment Selection. Prepared for the Office of Coal Research, U.S. Department of the Interior, College of Earth and Mineral Sciences, Pennsylvania State University, May 1973.

Battelle Memorial Institute. *Energy Perspectives.* Columbus, Ohio, 1973.

Bauer, Frederick L. "Wages in Bituminous Coal Mines," *Monthly Labor Review,* January 1968.

Berndt, Ernst R., and Wood, David O. "Technology, Prices, and the Derived Demand for Energy," *Review of Economics and Statistics* 57, August 1975.

Bowman, Mary Jean, and Haynes, W. Warren. *Resources and People in East Kentucky: Resources for the Future.* Baltimore, Md., 1963.

Carter, Richard D., and Dinkel, Janet S. "Economic Growth and Change in Kentucky, 1960-1970," *Economic Review,* October-November 1972.

Charles River Associates. *The Economic Impact of Public Policy on the Appalachian Coal Industry and the Regional Economy.* U.S. Department of Commerce, vols. 1, 2, 1973.

Coal Week, 11 August 1975.

Cook, Earl. "The Flow of Energy in an Industrial Society," *Scientific American,* September 1971, p. 137.

Edison Electric Institute. *Fuels for the Electric Utility Industry, 1971-1985.* New York, 1973.

Energy Alternatives: A Comparative Analysis. Prepared by the Science and Public Policy Program, University of Oklahoma, May 1975.

Energy Research and Development Administration. *Creating Energy Choices for the Future,* vol. 1. Washington, D.C., 1975.

Erickson, Edward W., and Spann, Robert M. "Supply Response in a Regulated Industry: The Case of Natural Gas," *Bell Journal of Economics and Management Science* 2, 1971.

Erickson, Edward W., and Waverman, Leonard, eds. *The Energy Question: An International Failure of Policy.* Vol. 2: *North America.* Toronto, 1974.

Exxon Company. *Energy Outlook, 1975-1990.* Houston, Texas, 1974.

Federal Energy Administration. *Project Independence: A Summary.* Washington, D.C., November 1974.

───── *Project Independence: Blueprint Final.* Prepared by the Interagency Task Force on Coal, Washington, D.C., November 1974.

Federal Power Commission. News Release of March 18, 1971. Washington, D.C.

───── *1970 National Power Survey,* pt. 1. Washington, D.C., 1971.

Federal Reserve System, Board of Governors. *Federal Reserve Bulletin,* September 1974.

Gilliam, Darrell, and Whitehead, David. *The Employment and Expenditures Effect on the Kentucky Coal Industry in 1970.* Lexington, Ky., 1972, p. 15.

Gordon, Richard L. *U.S. Coal and the Electric Power Industry.* Baltimore, Md., 1975.

Grayson, Leslie E., ed. *Economics of Energy: Readings on Environment, Resources, and Markets.* Princeton, N.J., 1975.

Greenbaum, M. E. *Kentucky Coal Reserves: Effects on Coal Industry Structure and Output.* Lexington, Ky., 1975.

Harvey, C. E. *Enforcing Weight Restrictions on Eastern Kentucky Roads.* Prepared for the Attorney General, Commonwealth of Kentucky, Office of Development Services and Business Research, University of Kentucky. Lexington, Ky., 1972.

───── *The Western Kentucky Coal Industry: An Economic Analysis.* Lexington, Ky., 1974.

Just, James E. *Changes in Energy Consumption.* Cambridge, Mass., 1973.

Kaysen, Carl, and Turner, Donald. *Antitrust Policy.* Cambridge, Mass., 1959.

Kentucky Department of Mines and Minerals. *Annual Report.* Lexington, Ky., 1969-1975.

Keystone Coal Industry Manual. New York, 1973, 1974.

Koch, James V. *Industrial Organization and Prices.* Englewood Cliffs, N.J., 1974.

Lichtblau, John H. "U.S. Demand for Atomic and Conventional Energy by 1982." Paper presented to the 39th Annual California Regional Fall Meeting of the Society of Petroleum Engineers, Los Angeles, 1967.

Mathematica, Inc. *Design of Surface Mining Systems in Eastern Kentucky.*

Prepared for Appalachian Regional Commission, Department of Natural Resources and Environmental Protection, Frankfort, January 1974.

The Mitre Corporation. *Impact of State Implementation Plans on Fossil Fuels Availability and Requirements.* Washington, D.C., August 1972, p. 50.

Morrison, Warren E. *An Energy Model for the United States, Featuring Energy Balances for the Years 1947 to 1965 and Projections and Forecasts to the Years 1980 to 2000.* Washington, D.C., 1968, 1970.

Moyer, Reed. "Competition and Performance in the Midwestern Coal Industry," *Journal of Industrial Economics,* July 1967.

────── *Competition in the Midwestern Coal Industry.* Cambridge, Mass., 1964, pp. 24-25.

National Coal Association. *Bituminous Coal Facts.* Washington, D.C., 1972.

────── *Steam-Electric Plant Factors.* Washington, D.C., 1967, 1973.

Nordhaus, William D. "The Allocation of Energy Resources," *Brookings Papers on Economic Activity* 3. Washington, D.C., 1973.

Resources Planning Associates. *Energy Supply/Demand Alternatives for the Appalachian Region.* Prepared for Appalachian Regional Commission. Washington, D.C., 1975.

Rieber, Michael. *Low-Sulfur Coal: A Revision of Reserve and Supply Estimates.* Urbana: Center for Advanced Computation, University of Illinois, November 1973.

Robert R. Nathan Associates, Inc. *Projections of the Consumption of Commodities Producible on the Public Lands of the United States, 1980-2000.* Prepared for the Public Land Law Review Commission. Washington, D.C., May 1968.

────── *A Review and Comparison of Selected United States Energy Forecasts.* Prepared for the Office of Science and Technology, Executive Office of the President. Washington, D.C., 1969.

Schweitzer, S. *The Limits to Kentucky Coal Output: A Short Term Analysis.* Lexington, Ky., 1973.

Tennessee Valley Authority. *Coal Reserves of Western Kentucky.* Chattanooga, 1969.

Texas Eastern Transmission Corporation. *Competition and Growth in American Energy Markets.* Houston, Texas, 1968.

U.S., Congress, House, Committee on Science and Astronautics, Subcommittee on Energy, 93d Cong. *Energy Facts.* Washington, D.C., 1975.

U.S., Congress, Senate, Council on Environmental Quality. *Coal Surface Mining and Reclamation.* Washington, D.C., 1973.

U.S. Department of Commerce. *Historical Statistics of the U.S., Colonial Times to 1957, and Statistical Abstract of the United States.* Washington, D.C., 1973.

U.S. Department of the Interior, Bureau of Mines, *Analysis of Tipple and Delivered Samples of Coal.* Report of Investigation 7997. Washington, D.C., 1975.

——— "Bituminous Coal and Lignite Distribution, Calendar Year 1972." Washington, D.C., 1973.

——— "Bituminous Coal and Lignite Distribution, Calendar Year 1973." Washington, D.C., 1974.

——— "Coal—Bituminous and Lignite," *Mineral Industry Surveys.* Washington, D.C., 1969-1974.

——— "Coal—Bituminous and Lignite," *Minerals Yearbook.* Washington, D.C., 1969-1974.

U.S. Department of the Interior, Bureau of Mines. "Comparative Transportation Costs of Supplying Low-Sulfur Fuels to Midwestern and Eastern Domestic Energy Markets," *Information Circular* 8614. Washington, D.C., 1973.

——— "Demonstrated Coal Reserve Base of the United States on January 1, 1974," *Mineral Industry Surveys.* Washington, D.C., 1974.

——— *Weekly Mineral Industry Surveys.* Washington, D.C., 1975.

U.S. Federal Trade Commission. *In the Matter of Kennecott Copper Corp., Finding of Fact, Conclusion and Final Order.* Docket 8765. Washington, D.C., May 1971.

Wall Street Journal, 15 March 1974, 5 April 1974.

Whitney, S. N. *Antitrust Policies,* vol. 1. New York, 1958.

Widner, Ralph R. "Current and Emerging Policies for the Extraction, Shipment and Use of Appalachian Coal," *Coal and Public Policy.* Center for Business and Economic Research, University of Tennessee, 1972.

Williams, Bobby A. "The Chinese Petroleum Industry: Growth and Prospects," *China: A Reassessment of the Economy.* A Compendium of Papers Submitted to the Joint Economic Committee, U.S. Congress, July 1975.

Index

acid pollution, 76
air pollution, 95
air-pollution standards, 34, 36-37, 46, 98
air-purity standards, 4, 11, 20, 29, 105
air-quality standards, 18, 29, 38, 144
Alabama, 17, 115, 117, 138, 144
American Electric Power, 89-90
anthracite coal, 47
Appalachia, 40
Appalachian Coal Basin, 14, 17, 48; output of, 42, 43, 44; price of coal from, 86
ash, 4, 17-18
Atomic Energy Commission, 30
atomic power, 154
auger mining: mine size, 55; productivity of, 78-80

Battelle Memorial Institute, 30
Bell County, 78
bituminous coal, distribution of, 21, 47, 105. See also coal
Boyd County, 60, 78
Breathitt County, 60, 78
breeder reactors, energy generation by, 30, 154, 155
British thermal units (BTU), 4; as a quality variable, 17-18, 21, 151; content of eastern Kentucky coal, 17; content of western Kentucky coal, 95
broad form deed, 73
Bureau of Mines. See United States Bureau of Mines

Canada, 37, 40-41
capacity use, 81
capital equipment, 10, 76, 85
captive mines, 12, 52-53, 88
captive output, 9
cartel. See Organization of Petroleum Exporting Countries
Carter County, 60
Charles River Associates (CRA), 35, 37, 41

China, 152
coal: as energy resource, 1, 151, 153-54; uses of, 11, 13; substitute for oil, 86
coal demand: function, 2; factors influencing, 2-4, 17; elasticities of, 3-4; retail, 40. See also electric utility industry
coal exportation, 40-41
coal market forms: spot market, 13; contract market, 13
Coal Mine Health and Safety Act, 43, 60, 68, 74, 156
coal-mining firms, structure of, 7
coal-oil mergers, effect on competition, 14
coal output: in Kentucky, 1, 14, 55, 60; in the United States, 6, 7, 99; concentration of, 9-10; of Kentucky surface mines, 14; of Kentucky underground mines, 14; in Appalachian Coal Basin, 42, 68; in eastern Kentucky, 64; in western Kentucky, 95-96, 98; in Eastern Interior Coal Basin, 120
coal reserves: in eastern Kentucky, 47-50; life of, 50, 119; in western Kentucky, 102-3
coal seams, 21; of eastern Kentucky, 14; of western Kentucky, 15; and productivity, 78, 81
coking coal, 3; usage and markets for, 11-12, 13, 37-39; European industry, 41
coking plants, 37, 105
Common Market, 41
competition, 7, 9-10, 14. See also concentration
concentration, 7, 9-10, 13
conglomerates, 13-14
Consolidation group, 7
conspiratorial behavior, 13
construction industry, and surface mining, 11, 55, 76
contract markets, 13, 51-52, 88
Council on Environmental Quality, 48

169

District 3, 33
District 6, 33
District 8, 21, 31, 33, 38-41
District of Columbia, 34
Dole, H. M., 154

Eastern Interior Coal Basin, 15, 33, 94, 95, 133; as supplier of coal, 105, 113-14, 119
eastern Kentucky, 14, 17; properties of coal from, 17-21; coal output of, 64; mine size distribution, 64, 66-72; productivity of, 73-76, 78-83, 85; price of coal from, 86-87, 89
economies of scale, 80, 133-35
electric-power demand, 22-24, 106
electric-power generation, 20, 24, 105, 110
Electric Power Research Institute (EPRI), 106-7
electric utility industry: demand for coal, 12, 21-22, 33-38, 105-7; consumption of coal, 15, 111-12, 115; consumption of oil, 20, 27, 36
Elliott County, 60
employment: in Kentucky coal industry, 14; in eastern Kentucky coal industry, 83; in western Kentucky coal industry, 95; in Eastern Interior Coal Basin, 100-102
energy crisis, 90
energy demand, 110, 156
energy/GNP ratio, 22-23
energy-resource inventory, 1, 151
energy shortage, 15
engineering cost estimates, 129
Environmental Protection Agency (EPA), standards and restrictions, 46-47, 117, 156
Europe, 37, 41
European Coal and Steel Community, 41

Federal Energy Administration (FEA), 106-7
Federal Power Commission (FPC), 29
Federal Trade Commission, 14
fixed carbon content, 90
Florida, 33, 115, 117, 138
Floyd County, 60, 78
Ford Foundation, 14
fossil fuels, 24, 27

"fuel cost adjustments," 36
fuel oil, 2, 24, 29. *See also* oil

gasification process, 20
gas plants, 105
Geological Service, 47
Georgia, 33, 115, 117, 138, 144
geothermal energy, 110, 155
Green River, 137
Greenup County, 60
Gross National Product (GNP), 22-24
Gulf of Mexico, 137

Haley, Ted D., 102
Harlan County, 60, 78
Henderson County, 102-3
Hopkins County, 102, 129

Illinois, 95, 117; producer concentration in, 13; coal output of, 98-99, 132-33; employment in, 100-101; as coal supplier, 111, 113, 119; mine size distribution, 125, 129, 133; transportation of coal from, 138-39; productivity in, 141-42; as consumer of western Kentucky coal, 144
Indiana, 95, 117; coal output of, 98-99, 132-33; employment in, 100-101; as coal supplier, 111, 113, 119; as consumer of western Kentucky coal, 115, 117, 144; mine size distribution, 125, 129, 133; transportation of coal from, 138-39; productivity in, 141-42
institutional environment, 70, 72
Iowa, 113

Jackson County, 60, 78
Japan, 37, 41

Kennecott Copper, 14
Kentucky: coal output of, 14; uses of coal from, 15; economic outlook of, 143
Knott County, 60
Knox County, 60

labor productivity: and reclamation requirements, 76; and mine size, 79-80; and surface mining, 125, 129-30
land reclamation. *See* reclamation

landslides, 76
Laurel County, 60
Lawrence County, 78
Lee County, 60
Leslie County, 78
Letcher County, 60, 78
lignite coal, 47
limestone scrubbers, 148-49
liquefaction process, 20
long-term contracts, 12-13, 51. *See also* contract markets
Louisville Gas and Electric Company, 94, 117

Martin County, 60, 78
Menifee County, 60
mergers, 14
metallurgical grade coal, 11
Michigan, 144
mine-mouth generating plants, 138
mines: number of, in Kentucky, 14; classification of size, 55, 62, 64
mine size, 64, 66-72, 79, 133
mining companies, 55
mining permits, 72-73
Minnesota, 138
Mississippi, 115, 117, 138
Mississippi River, 137
Missouri, 113
moisture, 90
monopoly power, 10
Montana, 113
Morgan County, 60, 78
mountain slopes, surface mining restrictions, 47-48
Moyer, Reed, 136-37
Muhlenberg County, 95, 102-3, 119, 129

National Economic Research Associates, 24
natural gas, 1, 29, 142, 151, 154; substitute for coal, 5, 12, 27, 144
New England states, 90-91
New York (state), 31
Nordhaus, William D., 153
Norfolk, 40
North American Coal Company, 12, 89-90
North Carolina, 33
nuclear power, 1, 30-31, 110, 151, 154-55

Ohio, 17; properties of coal from, 18, 49; transportation of coal from, 33, 138; coal output of, 43, 45, 48, 70; life of coal reserves in, 47, 50-51; mine size distribution, 64, 66-68, 72; strip mining productivity, 76, 78; mine operation, 81-83; price of coal from, 86-87, 88; as consumer of western Kentucky coal, 144
Ohio County, 102, 129
Ohio River, 137
oil, 1, 154; substitute for coal, 5, 12, 20, 29, 36, 144; use by electric utilities, 29, 36
oil embargo, 3, 20, 29, 86, 87, 91, 151
Organization of Petroleum Exporting Countries (OPEC), 1, 152-53
output. *See* coal output
overburden, 78, 81
Owsley County, 60, 78

Peabody Coal Company, 14, 120-21
Peabody group, 7
Pennsylvania, 17; properties of coal from, 18, 49; coal output of, 43, 44, 45, 48; coal reserves in, 47, 51; captive mines in, 52-53; mine size distribution, 64, 66-68, 71-72; strip mining productivity, 76, 78; mine operation, 81-83
per-capita income, 143
Perry County, 60
Persian Gulf, 153
petroleum, 151. *See also* oil
Petroleum Administration for Defense District 1, 29
pig iron, 3, 15, 37-39
Pike County, 55, 60, 78
Pittston Coal Company, 89
Poland, 91
pollution. *See* acid pollution; air pollution
pollution-control equipment, 117
price, 5-6; and coal output, 54-55; and productivity, 85-87, 139; of coal in Eastern Interior Coal Basin, 99-100; of coal and oil, 151-52
price controls, 86
price-quantity relationship, 3
price-supply relationship, 6
price tracking, 151

171

"process steam," 39
production: United States level, 7; Appalachian Coal Basin level, 42-44
production costs, 83, 85, 130-31, 135
productivity, 73-74, 76, 85; factors influencing, 80-81; and mine size, 125, 129; measure of, 139; of western Kentucky surface mines, 139, 141-42; of western Kentucky underground mines, 141
product-supply relationship, 6
profit maximization, 4-5
Project Independence, 153
public utility plants, 15. *See also* electric utility industry
purchase contracts, 12-13

railroads, use of coal by, 22
rail transportation, 135-37, 139
recessions, 91
reclamation: of Kentucky land, 14-15; requirements, 44, 45, 72, 76, 78, 88, 142
residual fuel oil, 27, 29
restrictive market behavior, 13
river barge, 33, 137-39
Rockcastle County, 60, 78
Rocky Mountain coal, 13, 113, 136, 148

safety equipment, 68, 75, 87
safety legislation, 75
scrap steel, 37-38
sedimentation, 76
shale oil, 151
shipping costs, 21
short-term marketplace. *See* spot market
solar energy, 110, 151, 155-56
South Carolina, 33
spot market, 13, 51-54, 81, 88
State Implementation Plan (SIP), 148
statistical cost analysis, 129
steam coal, 11
steel industry: use of coal by, 9, 11, 12, 15, 37-38, 91; and captive mines, 9, 51-52
strikes: in 1974, 7, 13, 86; in 1971, 82, 96, 98, 132-33
strip mining: mine size, 55; productivity of, 76, 78-80, 142
strip mining legislation, 87-88

subbituminous coal, 47
sulfur-cleansing equipment, 15, 20, 37, 50, 94
sulfur content, 4; of eastern Kentucky coal, 14, 17-18, 49-51, 90; of western Kentucky coal, 15, 95; and electric utility industry, 148-49
sulfur dioxide, 148-49
sulfur-emission standards, 15
surface mining: industry entry requirements, 10-11; growth of, 43-45, 72; mine size distribution, 55, 60, 64, 66-67; coal price trend, 87; employment in western Kentucky, 100-101; productivity of, 129-30, 142
survival technique, 129, 131

Tennessee, 17, 33, 106, 115, 117, 138, 144
Tennessee River, 137
Tennessee Valley Authority (TVA), 36, 102, 115, 117, 148-49
tidal energy, 110, 155
tipple and delivered samples, 95
topography, influence on mining operations, 15, 67-68
transportation: cost aspects of, 13, 17, 73, 113, 115, 135-39
truck transportation, 73, 138

underground mining: industry entry requirements, 10-11; decline of, 43-45; mine size distribution, 50, 60, 64, 66-67; productivity of, 78-79, 125, 129; coal price trend, 87; employment in western Kentucky, 100-101
Union County, 102
unions, 73, 82-83, 85. *See also* United Mine Workers
United Mine Workers, 6-7. *See also* strikes
United States: coal output of mines in, 6-7, 99; energy consumption in, 22, 153
United States Bureau of Mines, 47, 95-96, 102, 121, 123, 136, 137
United States Supreme Court, 14
unit train, 137, 139
University of Kentucky Institute for Mining and Minerals Research, 102
uranium, 30-31, 151

Virginia, 17, 31
volatile matter, 90

wages, 5, 83, 85
Washington, 113
Webster County, 102
western Kentucky: coal output of, 95-96, 98, 117, 119; employment in, 100-101; coal reserves in, 102-3, 119; markets for coal from, 115, 117; mine size distribution, 125; transportation of coal from, 137-39; productivity in, 141-42
Westinghouse Corporation, 110
Westmoreland Coal Company, 12, 89

West Virginia, 17; properties of coal from, 18, 49, 95; transportation of coal from, 40; coal output of, 43, 44, 45, 48; coal reserves in, 47, 50; captive mines in, 52-53; mine size distribution, 64, 67-72; strip mining productivity, 76, 78; mine operation, 81-83; labor strikes in, 96

Whitley County, 60, 78
Whitney, S. N., 9
wind power, 110
Wisconsin, 113, 115, 117, 144
Wolfe County, 60
Wyoming, 113

www.ingramcontent.com/pod-product-compliance
Lightning Source LLC
Chambersburg PA
CBHW032045150426
43194CB00006B/432